Shipbreaking: Hazards and Liabilities

Michael Galley

Shipbreaking: Hazards and Liabilities

Michael Galley
Law Research Centre
Southampton Solent University
Southampton
Hampshire
United Kingdom

ISBN 978-3-319-04698-3 ISBN 978-3-319-04699-0 (eBook)
DOI 10.1007/978-3-319-04699-0
Springer Cham Heidelberg New York Dordrecht London

Library of Congress Control Number: 2014944731

Printed on acid-free paper

Springer is part of Springer Science+Business Media (www.springer.com)

To the shipbreakers..

We shall not flag nor fail.
We shall go on to the end.
We shall scrap them on the beaches
We shall scrap them at the ocean's edge,
and on the intertidal zones.
We shall scrap them through fires and
explosions
We shall strip them of PCBs and asbestos,
We shall break them into pieces.
We shall never surrender.

(W.S.C. – almost)

Preface

Although the actual numbers may vary between sources, each year more than 700 ocean-going ships come to the end of their working lives and are scrapped, mainly on the beaches of the Indian sub-continent. Whilst the relative rankings of a dominant breaking state may also vary over time, the factors that many shipbreaking sites today have in common are the labour-intensive, largely unregulated, and highly polluting ways in which the ships are demolished, and the extensive, cumulative damage to both human health and the environment that has arisen from the process. Ships contain a wide range of hazardous materials, either incorporated into their structures or generated as operational wastes during their voyages. Introduced because of their inherent operational properties and also to comply with a range of international regulations, these hazards can remain largely inert until they are disturbed.

Many of the ships now being scrapped were built before the use of some of the hazardous materials employed in their construction was banned. The demolition process opens up and spreads these hazards around the thousands of workers involved in the breaking operation and around the environment in which they live and work. Until recently, most people were largely unaware of the pollution that was taking place. Shipowners (*i.e.* those disposing of vessels that they previously operated commercially) did little or nothing to pre-clean some of the hazards from their surplus ships before sending them to the breakers. As regulations became progressively enforced in one country, the shipbreaking industry would simply migrate to another, where restrictions were more relaxed, hazardous waste tending to follow the lines of least resistance. In recent years, this situation has been successfully brought to international attention and the emphasis has moved from *waste on ships* to *ships as waste*.[1]

The migration of the industry to developing states was characterised by Rousmaniere as one whereby occupational health risks moved from the developed states with mature infrastructures and appropriate capital and regulations to the

[1] Jones (2007).

largely rural areas of developing countries where such provisions are relatively weak. Furthermore, attempts to improve the economic, safety and social provisions of those engaged in shipbreaking are hampered by the fact that similar conditions may operate for most other sectors in these states.[2]

The risks and hazards involved in the demolition of ships are manifold, and include fires and explosions, falling from heights, being crushed by falling objects, etc. The focus here, however, is on the dangers arising from the release and handling of hazardous materials found onboard ships and the damage that they cause to the health of both humans and the environment. The aim of this work, which was originally the subject of a PhD research thesis, is to examine and to apportion liabilities of shipowners and shipbreakers for the safe removal of these hazardous materials from end-of-life ships under any relevant legal instruments. In addressing the subject, it considers a number of objectives, which relate to the various chapters beginning with a review of the history, locations and processes of the shipbreaking industry (Chap. 1). It follows with the role and mechanics of international law, especially with regard to dispute resolution, national sovereignty and the role played by Non-Government Organizations (NGOs) and the International Maritime Organization (IMO) in the formulation of law, and the extent to which existing national and international legislation has to date impacted upon the operation of the major shipbreaking industries (Chap. 2); particular attention is paid to the handling and transboundary movement of hazardous waste, including the contested applicability of the *Basel Convention* to end-of-life ships. Next, the extent applicability and application of existing legal instruments are examined from an international, regional (European) and national viewpoint (Chap. 3). The question of ship registration, particularly with regard to certain 'open' registers, is examined to determine the extent to which these might aid—and in a number of instances, positively encourage—anonymity of ownership and hence of liability (Chap. 4). The role of NGOs and the prompting of judgements from various national courts with regard to the application of international law are examined in a series of case studies where attempts (often successful) have been made by shipowners or shipbreakers to circumvent existing legal provisions (Chap. 5). Finally, it looks at the provisions of the new *Hong Kong Convention*[3] and certain perceived lacunae (Chap. 6), and at a number of other initiatives, both legal and commercial that have arisen to either promote the coming into force of the Convention or to operate independently, but in parallel with its provisions (Chap. 7). The extent to which these issues have been considered is summarised in the various sections of the final chapter (Chap. 8).

Since many legal instruments are based upon the legal definitions contained therein, it is appropriate that such a format be the basis for this book. Consequently, a number of personal definitions of the various aspects of liability, of shipbreaking

[2] Rousmaniere (2007).

[3] *The Hong Kong International Convention for the Safe and Environmentally Sound Recycling of Ships 2009.*

and of shipowners have been employed to express and define the viewpoint employed—not to indicate any specific bias (particularly, for example, in the use of 'shipbreaking'), but to define the rationale behind the terms and to maintain a consistency of approach.

In examining the question of liability for the final disposal of hazardous materials found onboard end-of-life ships, it is therefore important to define what is meant by the term 'liability' within the scope of this work. The question of **liability** is approached not in the more usual legalist manner of public liability and tort, nor in the financial sense of a charge that appears on a balance sheet (although ultimately both may be elements of the mischief occurring on the shipbreaking beaches), but in the somewhat broader realm of wider obligations, of moral or ethical liability, legal liability and practical liability.

In terms of shipbreaking, the **moral liability** of both shipowners and shipbreakers lies in their responsibilities, both singularly and jointly, to minimise the adverse impacts of the hazardous substances contained either within the structure or as operational wastes or cargo residues in the vessels that they consecutively own. In addition, shipbreakers have the obligation to provide a safe working environment for those employed on the actual demolition, and this may be extended back to shipowners (of however long a duration) to work with those yards that do operate in a safe manner and to avoid those with dubious reputations. The impacts of unsatisfactory operations fall upon the workers dismantling ships, especially those working on the beaches, upon the surrounding population, and the surrounding ecology. Medical care at the beaches has usually been basic, and compensation for injured workers minimal.

The timing and method of removal of hazardous material substances have been the subject of ongoing debate, largely initiated and developed by various NGOs. To date, both owners and breakers have generally been of the common opinion that the question of pre-cleaning ships prior to demolition basically be left unaddressed, the owners thereby maximising the financial gains from the sale of their ships, and the breakers similarly benefitting from the degree to which they choose to adopt (or not to adopt) a precautionary manner in their procedures.

Attempts to engage and encourage organizations to face this moral liability have been the focus of the NGOs' campaigning activities, and in this they have had some successes. As well as presenting their arguments at many relevant meetings of the IMO and shipping groups etc., Greenpeace and other members of the NGO Shipbreaking Platform have issued papers and critiques on many of the proposals of these organisations. An early target of the attentions of Greenpeace, the P&O Nedlloyd (subsequently Maersk) organisation subsequently formulated a policy of close association with selected Chinese breakers and supervised the demolition of its ships, and thereby was deemed to be an early model of ethical behaviour. Results within the actual shipbreaking industry have been less positive, the response to adverse publicity of the conditions prevailing being the classification of shipbreaking sites as restricted areas by state authorities. In an attempt to evade the liabilities that their ships represent, owners (and especially those of one-ship companies) have frequently sought to cover their ownership, not only through frequent changes of names and flags of the ships, but also through their own anonymity offered by various open registers; such actions may be said to aid the proliferation of sub-standard shipping.

Legal liability is addressed in this instance to the degree to which existing laws relating to shipbreaking operations are: firstly, formulated and directed to the industry; secondly, are observed by all parties to the shipbreaking process; and thirdly, are enforced in a manner that observes both the letter and the spirit of the legal instruments. The observance of legal provisions is incumbent upon all parties. Legal liability, in this context, is aimed at addressing the mischiefs existing and as such, often emanates directly from the moral liability towards the subject.

Some efforts have been made to reduce the impact of hazardous materials by banning their use under international laws during ship construction, but ships currently arriving at the breakers may still contain many materials employed before their prohibition came into force, and indeed banned materials are still finding their way aboard ships in the form of spares that may be sourced worldwide.

The question of the applicability of the *Basel Convention*[4] to shipbreaking has not only been one that has until recently been the subject of quite intense polarization of opinion, but has also been one that is still easily circumvented by shipowners, who find little difficulty in selling or reflagging their end-of-life ships just prior to final disposal, although here it is also important to recognise the difficulty in finding states included within the scope of *Basel* that are able to accommodate some of the larger vessels. Legal liability applies also to the processes and procedures as applied by the breakers, who so far have appeared to have paid little concern to the health and safety of the work force or to the environment. Similarly, the observance of legal obligations by national and local state authorities has, at times, been both inconsistent and highly subjective on the question of acceptance of various vessels arriving for demolition, and of observing the decisions of domestic apex courts on national standards defined for shipbreaking. The small but growing number of national courts' judgements are in agreement that pre-cleaning is the responsibility of the owner of the vessel.[5]

The moral and legal liabilities are, however, themselves subject to **practical liabilities**; what might be done, or what should be done, is limited by physical realities of the situation prevailing. A primary example of this, and a problem highlighted by those opposing current practices, is the physical problem of fully pre-cleaning or decontaminating a vessel of its inherent hazards prior to its departure to the breaker's yard. Whilst such an exercise might significantly reduce the hazard levels for those actually undertaking the demolition, it is most likely to leave a ship in a state that is deemed unseaworthy and able to progress to the breakers

[4] *Basel Convention on the Control of Transboundary Movements of Hazardous Wastes and their Disposal* 1989.

[5] See Chap. 5, Case Studies. Where, due to the anonymity provisions of various flag states, the identification of the beneficial owner has been impossible to determine, the exporting state itself has sometimes had to bear the cost of pre-cleaning or demolition, *e.g.* the *Sandrien* and *Sea Beirut*.

under its own power, without resorting to long distance towage—a task that is not only expensive and risky, but also highly problematic in that the required fleet of ocean-going tugs no longer exists. In practice, pre-cleaning may actually be regarded as the initial stage of demolition.

From a shipbreaker's point of view, and especially in the sub-continent, the high cost of money necessary to buy ships results in a pressure to complete the demolition in as fast, yet as cheap a manner as possible, using labour that appears to be easily replaced. Many of these shipbreaking operations are operated not as public companies but as private family businesses; hence shareholder considerations are not of significance, yet the value of money may limit the value placed upon considerations of health and safety. In fact this (subjective) responsibility is answerable—and is being answered—in the form of a reduction in prices received by owners from certain yards that do practise more responsible methods, a penalty that a small but growing number of shipowners appear willing to accept in the name of public image. In addition, other practical considerations, such as the limited number of high tides available for beaching, the imposition of various import and sales taxes and customs duties, demands from local construction industries for the products of shipbreaking, together with monsoons, religious festivals and the intense competition from other operators (both domestic and international), all serve to impose a practical liability upon shipbreakers to perform their activities in the quickest and easiest (and hence cheapest) manner possible.

Added to this, the fluctuations in legal judgements from domestic courts place demands upon shipbreakers that appear to be at times unpredictable, and which can slow, halt or even close shipbreaking yards—a prime example of this concerned the closure of the shipbreaking yards of Bangladesh over a period of 2 years, whilst the shipbreakers, the NGOs, the courts and the government ministries all fought for control of, and revisions to, the industry—see Chap. 3.

At this juncture, it is necessary to consider nomenclature. Traditionally known as '**shipbreaking**', the industry is now referred to by a range of names, which have attracted both political and interpretive associations to their individual use. 'Shipbreaking' remains the term in use with the International Labour Organization (ILO), and with environmental NGOs such as Greenpeace and the Basel Action Network (BAN) and carries with it, perhaps, the image of a basic heavy and traditional industry. The term is also preserved in such titles as the Pakistan Ship-Breakers Association, the Bangladesh Shipbreakers' Association and the Iron Steel Scrap and Shipbreakers Association of India. 'Ship dismantling' has been the preferred—perhaps more neutral—term of the Basel Convention (BC), and by the European Commission (EC), whilst 'scrapping' has currency with shipowners and with the Joint Working Group (JWG) at the International Maritime Organization (IMO), consisting of the IMO, the ILO, and the Basel Convention (BC). 'Demolition' or 'demo' is often used by brokers, whilst 'disposals' often appears in shipping statistics, both words perhaps indicating something that is little more than another commodity in the world of financial trading. 'Ship recycling', however, is the term that has become of more widespread use of late and is the handle now favoured by the IMO, the shipping industry in general and, increasingly, by the shipbreakers themselves. The term has also been adopted by the EC in its new *Ship Recycling Regulation* (*SRR*). Its use is intended to project, albeit in somewhat anodyne terms, the 'greener' message that ships-for-disposal are transformed into products that have either a direct or indirect further and useful life in other forms or applications, rather than being merely the subject of a highly dangerous and polluting industry that they also are.

As will be elaborated below, not only does the scrap metal form a very valuable resource for the shipbreaking nations involved, but a high proportion of a ship's equipment finds its way back into reuse in shore-based industries and commercial and domestic addresses at a level that is not experienced in the Western nations. This enhanced level of reuse, in a sense, places the activity even higher up the European waste hierarchy than mere recycling.

The overall process may be divided into two distinct phases—the actual scrapping of the ships and the recycling of the resultant scrap. The scrap metal is transported from the breaking sites to rolling mills for recycling, often for conversion into low grade reinforcing rods (rebar) for use by the local construction industries. Here we are not concerned with the latter phase, but purely with the breaking or scrapping operation; for this reason, the term 'shipbreaking' will be used throughout in preference to 'ship recycling'. For that same reason, the new *Hong Kong International Convention for the Safe and Environmentally Sound Recycling of Ships* 2009 (*HKC*) might also be more appropriately named the *Hong Kong International Convention for the Safe Breaking* (*or Scrapping*) *of Ships*, since it also is restricted purely to the process of demolition and does not cover the actual recycling process—which takes place beyond the breakers' sites. Neither does the proposed European *SRR*,[6] whose definition of recycling specifically excludes the actual recycling operation. The legalities of definitions aside, however, the term 'recycling' was defended by an IMO staff member as one that promotes '*a constructive and productive activity*'.[7]

The definition of '**shipowner**' is a somewhat more flexible issue, at least in the context here. In the main, it will be used to define the last owner of a ship who employed it for traditional trading purposes and who, therefore, holds all responsibilities for the ship, its contents and its disposal. As a ship may pass through the hands of several owners, including cash buyers, just prior to scrapping, so too should the associated liabilities, although this has often been questioned in terms of those who may exercise ownership for a matter of days or even hours. The new *HKC* and the EU's *SRR* both recognise these short-term owners as owners in the fullest sense.

In terms of national legislation, especially of the shipbreaking states, the definition of ownership may also be taken to mean or include the final buyer who

[6] Proposal for a Council Decision requiring member States to ratify or to accede to the Hong Kong International Convention for the safe and environmentally sound recycling of ships, 2009, in the interests of the European Union European Commission 2012.

[7] Discussions with IMO representative, 1 December 2009.

procures a vessel for demolition, *i.e*. the operator of a shipbreaking yard or plot. The distinction is important, especially when considering the liability for the safe removal of hazardous materials found on board; in all instances, that liability lies with the 'owner,' but whether this is the ultimate owner—the breaker—or any that comes before, appears to be problematic. Traditionally, the responsibility for pre-cleaning (prior to demolition) has been laid at the feet of what might be termed the 'exporter' of the ship not only by the campaigning NGOs, but also by a number of judgements from national courts—see Chap. 5. However, current practices, with the exception of a small but growing number of shipowners, appears to leave all owners (at whatever stage in a ship's disposal) in agreement that it is both practical and economically advantageous for the matter of cleaning prior to disposal to be left with the breaker, thereby obviating costs for the exporter, and lowering the price for the breaker.

Attempts to introduce voluntary standards of operation on the breaking beaches have so far proved to be ineffective, since the voluntary codes devised are rarely accompanied by monitoring, enforcement and penalties. International and regional legal instruments on the transhipment of hazardous waste have been strongly resisted by many groups, including ship owners and the owners of shipbreaking operations, as being inapplicable, and are easily circumvented with regard to their applicability to ships-for-scrap. The *HKC* is an attempt by the IMO to bring order, control and improvements to the breaking industry, but although its measures are mandatory, at least upon those states that ratify it, it too is based upon guidelines that are open to national interpretation, and specific sanctions do not appear to be included.

References

Jones SL (2007) A toxic trade: shipbreaking in China. A China environmental health project fact sheet. Western Kentucky University, Bowling Green

Rousmaniere P (2007) Shipbreaking in the developing world. Int J Occup Environ Health 13:359

Legislation Cited (Including Statutes & Declarations)

International

1907: Hague Convention on the pacific settlement of international disputes
1945: Charter of the United Nations
1945: Statute of the International Court of Justice
1948: Convention on the Inter-Government Maritime Consultative Organization
1954: London Dumping Convention and the 1996 London Protocol
1954: International Convention for the prevention of pollution of the sea by oil (OILPOL)
1958: Geneva Convention on the high seas
1959: Antarctic Treaty and the 1991 Madrid Protocol
1966: International Convention on load lines
1967: Treaty on principles governing the activities of states in the exploration and use of outer space, including the Moon and other celestial bodies
1969: Vienna Convention on the law of treaties
1969: International Convention on civil liability for oil pollution damage
1972: Declaration of the Stockholm Conference on the human environment (UNCHE)
1972: London Convention on the prevention of marine pollution by dumping of wastes and other matter
1973: International Convention on the prevention of pollution from ships, and the 1978 London Protocol (MARPOL)
1973: Convention on the international trade in endangered species of wild fauna and flora (CITES)
1974: International Convention on the safety of life at sea (SOLAS)
1979: Geneva Convention on long-range transboundary air pollution
1982: United Nations Convention on the law of the sea (UNCLOS)
1982: World Charter for nature
1985: Vienna Convention for the protection of the ozone layer and the 1987 Montreal Protocol
1986: United Nations Convention for registration of ships (not yet in force)

1987: Montreal Protocol on substances that delete the ozone layer (Protocol to the Vienna Convention for the protection of the ozone layer)
1989: Basel Convention on the control of transboundary movements of hazardous wastes and their disposal
1992: Rio Declaration on environment and development
1992: Convention on biological diversity
1992: UN Framework Convention on climate change and the 1997 Kyoto Protocol
1998: Rome Statute of the International Criminal Court
1998: Rotterdam Convention on the prior informed consent procedure for certain hazardous chemicals and pesticides in international trade
2001: Stockholm Convention on persistent organic pollutants (POPs)
2001: International Convention on control of harmful anti-fouling systems for ships
2001: Articles of Responsibility for internationally wrongful acts (ARSIWA) (ILC)
2004: International Convention for the control and management of ships' ballast water and sediments (not yet in force)
2007: Nairobi International Convention on the removal of wrecks (not yet in force)
2009: Hong Kong International convention for the environmentally sound recycling of ships (not yet in force)

EU

1975: Council Directive 75/442/EEC of 15 July 1975 on waste
1984: Council Directive 84/631/EEC of 6 December 1974 on the supervision and control within the European Community of the transfrontier shipment of waste
1987: Council Directive 87/217/EEC of 19 March 1987 on the prevention and reduction of environmental pollution by asbestos
1991: Council Regulation (EEC) 613/91 of 4 March 1991 on the transfer of ships from one register to another within the Community
1993: Council Regulation (EEC) 259/93 of 1 February 1993 on the supervision and control of shipments of wastes within, into and out of the European Community
1995: Council Directive 95/21/EC of 19 June 1995 concerning the enforcement in respect of shipping using Community ports and sailing in the waters under the jurisdiction of the Member States, of international standards for ship safety, pollution prevention and shipboard living and working conditions (port state control)
2001: Directive 2001/106/EC of the European Parliament and Council of 19 December 2001 amending Council Directive 95/21/EC
2003: Regulation (EC) 782/2003 of the European Parliament and of the Council of 14 April 2003 on the prohibition of organotin compounds on ships
2006: Directive 2006/12/EC of the European Parliament and of the Council of 5 April 2006 on waste (the Waste Framework Directive)

2006: Regulation (EC) 1013/2006 of the European Parliament and of the Council of 14 June 2006 on shipments of waste
2008: European Parliament Resolution of 21 May 2008 on the Green Paper on better ship dismantling (2007/2279(INI))
2009: Directive 2009/16/EC of the European Parliament and the Council of 23 April 2009 on port state control
2009: Commission Regulation (EC) 967/2009 of 15 October 2009 amending Regulation (EC) 1418/2007 concerning the export for recovery of certain waste to certain non-OECD countries
2009: Directive 2009/15/EC of the European Parliament and of the Council of 23 April 2009 on common rules and standards for ship inspection and survey organizations
2010: Commission Regulation (EU) 757/2010 of 24 August 2010 amending Regulation (EC) No 850/2004 of the European Parliament and of the Council on persistent organic pollutants as regards Annexes I and III
2013: Ship Recycling Regulation (not yet in force)

Other

UK

1994: Transfrontier shipment of waste Regulations SI 1994/1137.
2007: Transfrontier shipment of waste Regulations SI 2007/1711
2007: Environmental Permitting (England and Wales) Regulations SI 2007/3538 (as amended)

Bangladesh

1974: The Territorial water and maritime zone Act
1988: Basel Convention Act
1989: Marine and fisheries Ordinance
1995: Environment protection Act
1997: Environment protection Rules
2006: Labour law Act
2011: Ship breaking and ship recycling Rules 2011
2011: Hazardous waste and ship breaking management Rules

India

1926: Trades union Act
1947: Industrial disputes Act
1948: Factories Act (as amended 1987)
1958: Merchant shipping Act
1979: Interstate migrant workmen (regulation of conditions of service) Act
1981: Gujarat Maritime Board Act
1986: Environment protection Act
1989: Hazardous waste (management and handling) Rules (amended 2000)
2000: Gujarat Maritime Board (prevention of fire & accidents for safety & welfare of workers and protection of environment during ship breaking activities) Regulations
2003: Gujarat Maritime Board ship recycling Regulations
2003: Hazard waste (management and handling) amendment Rules
2006: Gujarat Maritime Board (conditions and procedures for granting permission for utilizing ship recycling plots) Regulations
2006: Labour law Act

Netherlands

1992: Environmental management Act
1994: General administrative law Act
2004: Environmental management Act

Turkey

1970: Regulation for Aliağa shipbreaking yards
1974: Regulation on workers' health and occupational safety
1983: Environmental Law no. 2872
1991: Regulation on solid waste control (amended 2005)
1993: Regulation on hazardous chemicals
1995: Regulation on the control of hazardous wastes
1997: Regulation for Aliağa shipbreaking yards
2001: Solid wastes control regulation (amended 2001)
2008: Regulation on general principles of waste management

USA

1976: Toxic substances control Act
1990: Oil pollution Act

Case Law Cited

International Court – ICJ

1977: *Gabičkovo-Nagymaros Project case* (*Hungary v Slovakia*) 1997. ICJ Reports, 1997

International Court – ECJ

1990: Joined cases *Vesso and Zaneth* case C-206-207/88 28.3.1990, *Zanetti* case C-389/88 28.3.1990. [1990] ECR I-1461

1991: *R v Secretary of State for Transport ex parte Factorama*, [1991] Case C-221/458 89 ECR I-3905

1992: *Commission v. Ireland* [1992] Case C-280/89, ECR I-6185

1992: *Anklgemyndigheden v. Poulsen and Diva Navigation* [1992] Case 286/90, ECR I-6019

1997: *Commission v. Hellenic Republic* [1997] Case C-62/96, ECR I-6725

1997: *Arco Chemie Nederland* and others Joined cases C-418/97 and C-419/97 [2000] ECR I-7411

2000: *Palin Granit Oy* Case C-9/00 [2002] ECR I-3533

2000: Joined cases *Oliehandel Koeweit BV and others*, Case C-307/00; *Slibverwerking Noord-Brabant NV*, *Glückauf Sondershausen Entwicklungs- und Sicherrungsgesellschaft mbH*, case C-308/00; *PPG Industries Fiber Glass BV*, case C-309/00; *Stork Veco BV*, case C-310/00; *Sturing Afvalverwijdering Nord-Brabant NV*, *Afverbranding Zuid Nederland NV*, *Mineralplus Gesellschaft für Mineralstoffaufbereitung und Verwertung GmbH*, *formerly UTR Umwelt Gmbh v. Minister van Volkshuisvesting, Ruimtelijke Ordening en Milieubeheer*, case C-311/00 [2003]

2001: *Abfall Service AG (ASA) v Bundesminister für Umwelt, Jugend und Familie* Case no 6/01 [2003]

2003: *Paul van de Walle and others v Texaco Belgium SA* Case C-1/03 [2004] ECR I-07613

UK

1991: *R v Swale Borough Council, ex parte Royal Society for the Protection of Birds* [1991] 1 PLR 6
1994: *R v Inspectorate of Pollution, ex parte Greenpeace Ltd* (*No.2*) [1994] 4 All ER 329
1997: *R v. Bolden and Dean* (*The Battlestar*) [1997] 2 Int. M.L
1997: *R v Secretary of State for the Environment ex parte Royal Society for the Protection of Birds* (*Lappell Bank case*) [1997] Env LR 431

France

2006: Le Conseil d'État 288801 15.2.2006. *Judgement on the Clemenceau*

India

1995: *Research Foundation for Science Technology National Resource Policy v. Union of India*
1996: *Vellore Citizens' Welfare Forum v. Union of India* (1996) 5 SCC 647
1997: *People's Union for Civil Liberties v. Union of India* (1997) 3 SCC 433
1999: *A.P. Pollution Control Board v. Prof. M.V. Nayudu* (1999) 2 SCC 718
2002: *T.N. Godavarman Thirumalpad v. Union of India and Ors.* (2002) 10 SCC 606
2003: Supreme Court of India. *Order on Civil Original Jurisdiction Writ Petition No. 657 of 1995*
2005: Research Foundation for Science Technology National Resource Policy v. Union of India (2005) 10 SCC 510
2012: Supreme Court of India Civil Original Jurisdiction I.A. Nos. 61 and 62 of 2012 in Writ petition (C) N0. 657 05 1995 Research Foundation for Science Technology and Natural Resource Policy v. Union of India and Ors

Netherland

2002: Council of State, The Hague, *Upperton Ltd. v the Minister of Housing, Spatial Planning and the Environment*. LJN number AE4310 Case number 200105168/2

2007: Dutch Council of State, 2007. *Decision on the Otapan* Case No. 2200606331/1211.2.07. The Hague, Netherlands

USA

1982: *US v. Marino-Garcia* 1982 67 9 F 2d 1373 (11th Cir.1982)

2009: *United States Environmental Protection Agency*, Region 9, San Francisco, Consent Agreement and Final Order Docket no. TSCA-09-2008-0003 relating to violation of 40 C.F.R.§761.97 and section 15(1) of TSCA, 15 U.S.C.§2614(1) by exporting the Oceanic containing PCBs and PCB items for disposal outside the United States

2011: *Basel Action Network and Sierra Club v US Environmental Agency* [EPA–HQ–OW–2013–0157; FRL—9798–2]

Turkey

2003: Izmir, Turkey 2nd Administrative Court 2003. Case no. 2002/496 (The *Sea Beirut case*) Decision no. 2003/1184

Palestine

1948: *Naim-Molvan v. Attorney-General for Palestine* [1948] A.C.351

Acknowledgments

During the course of this research, opinions and comments were obtained from numerous individuals who represented a cross-section of interests involved, from shipbreaking representatives to those involved in formulating legislation and those responsible for its enforcement. Their input and guidance were much appreciated and instrumental in guiding the final output, especially that of Roy Watkinson, who bore my questions throughout the research with great patience. My thanks also to those organizations who gave me permission to quote from their own publications, websites, etc. The IMO request that I add that 'the quoted material may not be a complete and accurate version of the original material and the original material may have subsequently been amended'.

I should particularly like to give thanks to my friends Chrissie and Trevor Lawson, who sustained me in many ways during the long and regular commutes to University; to the staff of the University's Mountbatten Library (heroes all), especially their law specialist Hanna Young; and above all to my Director of Studies Professor Patricia D. Park, founder of the Law Research Centre at Southampton Solent University, and *Supervisor Extraordinaire*, who worked hard to keep me on the academic straight-and-narrow whilst making research such a pleasure.

Southampton, UK
November 2013

Michael Galley

Abbreviations

ABP	Associated British Ports
ABS	American Bureau of Shipping classification society
ACM	Asbestos-containing material
AERB	Atomic Energy Regulatory Board (India)
ARSIWA	Articles of Responsibility for Internationally Wrongful Acts 2001 (ILC)
ASSBY	Alang and Sosiya Shipbreaking Yard
BAN	Basel Action Network
BARC	Bhabha Atomic Research Center (India)
BC	Basel Convention
BELA	Bangladesh Environmental Lawyers Association
BIMCO	Baltic and International Maritime Council. Trade organization representing ship owners, brokers, agents and others
BSBA	Bangladesh Ship Breakers' Association
CA	Competent Authority
CBR	Central Board of Revenue (Pakistan)
CETP	Combined effluent treatment plant
CFC	Chloroflurocarbons
COP	Conference of Parties
CPA	Chittagong Port Authority (Bangladesh)
DASR	Document of Authorization to conduct Ship Recycling
DEFRA	Department for Environmental and Rural Affairs (UK)
DEMOLISHCON	BIMCO's standard contract for the sale of a vessel for scrapping
DFDS	Det Forenede Dampskibs-Selskab (Danish)—The United Steamship Company
DfT	Department for Transport
DNV	Det Norske Veritas
DWT	Deadweight tonnage—the carrying capacity of a ship when fully loaded. Includes cargo, bunkers, water (boilers, ballast and potable), stores, passengers and crew

EA	Environment Agency (UK)
EC	European Community
ECHR	European Court of Human Rights
ECJ	European Court of Justice
ECOSOC	UN Economic and Social Council
EEZ	Economic Exclusion Zone
EMSA	European Maritime Safety Agency
EPA	US Environmental Protection Agency
FIDH	International Federation of Human Rights
FNV	Federati Nederlandse Vakbeveging - a federation of Netherland Trades Union
GATT	General Agreement on Tariffs and Trade
GEPIL	Gujarat Enviro Protection and Infrastructure Ltd.
GL	Germanischer Lloyd classification society
GMB	Gujarat Maritime Board (India)
GMS	Global Marketing Systems (cash buyer)
GRI	Global Recycling Initiatives
GSL	Global Shipping LLC
GSSDF	Green and safe ship dismantling facility
GT	Gross tonnage. Internal capacity of a ship measured in units of 100 cu. ft.
GPCB	Gujarat Pollution Control Board (India)
HPC	High Power Committee (India)
HSD	Hariyana Ship Demolitions Pvt. (India)
IA	Intervention Application (India)
IBC	International Business Corporation
ICAM	Integrated Coastal Area Management
ICZM	Integrated Coastal Zone Management (EU)
IGO	Intergovernmental Organization
ICIHM	International Certificate on Inventory of Hazardous Materials
IHM	Inventory of Hazardous Materials
ILC	International Law Commission (UN)
ILO	International Labour Organization (UN)
IMF	International Monetary Fund
IMO	International Maritime Organization (UN)
INTERCARGO	International Association of Dry Cargo Ship Owners
INTERTANKO	International Association of Independent Tanker Owners and Operators of Oil and Chemical Tankers
IRIN	Integrated Regional Networks (UN)
IRRC	International Ready for Recycling Certificate
ISRA	International Ship Recycling Association
ITF	International Transport Workers Federation
ITOPF	International Tanker Owners Pollution Federation Ltd.
JWG	Joint Working Group

IPOS	Indian Platform on Ship Breaking
LDC	London Dumping Convention 1954
LDT	Light displacement tonnage—the actual weight of the ship, excluding cargo, stores, fuel, ballast, passengers and crew and used to calculate its scrap value
LR	Lloyds Register of Shipping classification bureau
MARAD	US Maritime Administration
MARPOL	International convention for the prevention of pollution from ships 1973 as modified by the 1978 Protocol (MARPOL 73/78)
MCA	Maritime and Coastguard Agency (UK)
MEA	Multilateral Environmental Agreement
MEPC	Marine Environmental Protection Committee (IMO)
MoA	Memorandum of Agreement
MoEF	Ministry of Environment and Forests (India)
MoU	Memorandum of Understanding
MPT	Mumbai Port Trust
NCL	Norwegian Cruise Lines
NGO	Non-Government Organization
NIOH	National Institute of Occupational Health (India)
NK	Nippon Kaiji Kyokai (Japanese classification society)
NORAD	Norwegian Agency for Development Co-operation
OCHA	Office for the Co-ordination of Humanitarian Affairs (UN)
OCIMF	Oil Companies International Marine Forum
OEWG	Open-ended Working Group
PAH	Polycyclic Aromatic Hydrocarbons
Panamax	Largest size of ship able to transit the Panama Canal. Typically 65,000–85,000 DWT
PCB	Polychlorinated Biphenyls
PCIJ	Permanent Court of International Justice
PIC	Prior Informed Consent (Art. 6, *Basel Convention*)
POP	Persistent Organic Pollutants
PIL	Pacific International Lines, Singapore
PISC	Platinum Investment Services Corp. based in Monrovia, Liberia.
PPE	Personal protection equipment
PBI	Priya Blue Industries Pvt. Ltd. (India)
PVC	Polyvinyl chloride
RINA	Royal Institute of Naval Architects. Also Registro Italiane Navale classification society
Ro–ro	Roll-on, roll-off vehicle ferries
SBC	Secretariat of the Basel Convention
SCMC	Supreme Court Monitoring Committee on Hazardous Waste (India)
SDI	Ship Decommissioning Industries Corporation

SOLAS	International Convention for the Safety of Life at Sea 196.
SOSREP	Secretary of State's Representative (UK)
SPCB	State Pollution Control Board (India)
SRO	Statutory Regulatory Order (Bangladesh)
SRF	Ship Recycling Facility
SRP	Ship Recycling Plan
SRR	(European) Ship Recycling Regulation
SYBAss	Superyacht Builders Association
TBT	Tributyl tin
TEC	Technical Experts Committee on Management of Hazardous Wastes relating to Ship breaking (India)
TSCA	Toxic Substances Control Act (USA)
ULCC	Ultra Large Crude Carrier. Tanker of 320,000 DWT and above
UN	United Nations
UNCLOS	United Nations Convention on the Law of the Sea 198.
UNCTAD	United Nations Conference on Trade and Development
UNEP	United Nations Environmental Programme
UNESCO	United Nations Educational, Scientific and Cultural Organization
UNIDO	United Nations Industrial Development Organisation
VCLT	Vienna Convention on the Law of Treaties 196.
VLCC	Very Large Crude Carrier. Tanker of 200,000–320,000 DWT
WSR	(European) Waste Shipment Regulation
WTO	World Trade Organisation
YPSA	Young Power in Social Action (Bangladesh)

Contents

List of Figures

List of Tables

Frontpiece Aerial view of Alang 3.11.2009. Note the discharge trails and the oil sheen on the sea

Chapter 1
Industry Development and the Process of Disposal

1.1 Background

Whilst ships are designed, constructed and operated in an environment of high technology and legislative control, their disposal is usually undertaken in a distinctly opposing environment of low cost, low technology, high labour content, high risk and minimal, or minimally enforced, legislation. The dangers inherent in shipbreaking are manifold. In addition to the dangers from hazardous materials, the workers are at risk from fires and explosions; falling objects; falling from heights; oxygen deficiency in confined spaces; biological hazards; long working hours and the burdens of insecure employment; poverty; low wages; inadequate housing and sanitation; inadequate accident prevention and inspection procedures etc. The International Metalworkers Federation now rates shipbreaking as one of the world's most dangerous occupations.[1] Such is the nature of the industry that on December 12th each year the Young Power in Social Action Group organises events to commemorate those who have died in the industry and publishes a list of the dead.[2]

For many years, shipbreaking has been an industry that has not operated in the forefront of public perception. Unwanted wooden ships could merely be abandoned and left to rot naturally. With the advent of metal ships, abandonment remained (and still is) an option for owners of obsolete vessels but, in the main, ships were often scrapped in or close to the sites at which they had been built. Even large ships, such as the UK's fleets of capital ships sent for disposal after the two World Wars, were quietly broken up, in a myriad of sites around the coast, the exceptions to this general anonymity being some of the more celebrated liners that ended their lives amid much public recognition. Final demolition was often undertaken in a manner somewhat similar to that operated on the beaches of the Indian sub-continent today, with hulls, lightened by the removal of engines and upper works, being progressively towed inshore for final demolition on the shore or river bed at low tide.

[1] International Metalworkers Federation (2003).

[2] Young Power in Social Action (2007).

M. Galley, *Shipbreaking: Hazards and Liabilities*, DOI 10.1007/978-3-319-04699-0_1,

International shipping carries some 90 % of the world's trade and has become a distinctly global industry. Of the 50,000+ merchant ships in the world fleet,[3] some 700 plus are deemed to be surplus to requirements and sent to breakers' yards each year to be demolished. From the breakers' sites, the resultant scrap is usually delivered to re-rolling mills for conversion into rebar for the local building industries. Some 80–90 % of a ship's 'dry' weight represents reusable or recyclable materials—mainly steel—depending upon the type of ship.[4]

When looking at the state of the shipbreaking industry, it is also necessary to consider the workings of the shipping industry itself, which is the supplier to the various breakers' yards. The break between the two is blurred by the actions of intermediaries between the owners and the breakers, intermediate brokers and cash buyers, who arrange the final disposal of surplus ships. Ownership can, and often does, change hands several times between the final operator and the shipbreakers, and this ability to change registrations and names can make the establishing of ownership difficult. This difficulty is further aided by the use of certain open registers, which decidedly assist in protecting, even promoting, the anonymity of ownership—see Chap. 4.

The campaign to reform and improve the standards of shipbreaking spread to international levels and resulted in the growth of a political momentum, a number of judgements from various national courts prohibiting the release of 'contaminated' ships for scrapping[5] and, in 2009, the adoption of the *Kong International Convention for the Safe and Environmentally Friendly Recycling of Ships* (*HKC*). Although it is acknowledged that there have been distinct improvements in sections of the industry over recent years, there is a focus on examining the more traditional practices of shipbreaking, since it is these practices that gave rise to the accumulated pollution of the surrounding environments and to the call for international remedial action that in its turn led to the adoption of the *Hong Kong Convention*.

[3] As at 31.10.2010, International Chamber of Shipping (2012).

[4] Andersen (2001), p. 8.

[5] Although there have been a (small) number of cases in which ships have been demolished or cleaned under strict conditions at sites within the EU—examples include the *Clemenceau* (scrapped in Hartlepool), the *Sandrien* (scrapped by the port of Rotterdam), and the *Otopan* (pre-cleaned in Rotterdam)—other high profile ships such as the *Blue Lady*, and the *Riky*, have finished up on the beaches of the sub-continent, despite legal action from home or exporting states—see Chap. 5.

1.2 Ship Disposal

1.2.1 Decommissioning and Selling

Once it is determined that a ship is surplus to requirements, the owner has a number of options as to how best to manage an expensive, but depreciating asset. If trading opportunities allow, the most frequent and favourable course of action is probably to sell the vessel to a new owner for further trading, or perhaps for conversion to an alternative use, since the value of a vessel sold for further trading may be twice that of a vessel sold for scrap.[6] In times of adverse trading conditions, it may be deemed prudent to put the vessel into lay-up, whether hot or cold lay-up,[7] if it is believed that trading conditions will improve within a foreseeable timescale. But lay-up is expensive to maintain and can represent a high level of cost—especially on re-commissioning—for a vessel that is generating no effective returns.[8] It may be determined that a vessel no longer has an economic trading life, since increasing costs of operations, surveys, repairs and insurance may prove to be prohibitive. Technical or legal obsolescence may also become a barrier to continued operations due to the high costs of retrofitting vessels, especially passenger vessels. The costs of retrofitting to meet the new *SOLAS*[9] requirements of 2010 resulted in many withdrawals from service; similarly, the costs of retrofits necessary to meet the new legislative requirements of ballast water and carbon emissions legislation may have similar impacts. In all these instances, vessels may be offered for scrapping on decommissioning.[10] Restrictions in the supply of credit to either owners or breakers may also have a major impact, either positive or negative, upon the level of activity in the breakers' yards. In this way, the shipbreaking market creates a balance

---

[6] International Federation of Human Rights (2012), p. 9.

[7] In hot lay-up (for periods of up to 12 months) the ship is maintained by a reduced crew, and machinery is kept in an operational state. Re-commissioning may take 1 week. In cold lay-up (for periods up to 5 years) machinery is taken out of service and the ship is electrically dead apart from emergency power. Manning is at a minimum and the lay-up site is usually remote. Re-commissioning may take up to 3 months. Det Norske Veritas (2009).

[8] The extreme case of lay-up is probably represented by the US National Defense Reserve Fleet, which accumulated due to the moratorium on exporting ex-naval and naval auxiliary vessels overseas for scrapping. At its peak the Reserve Fleet numbered almost 500 vessels although numbers are now greatly reduced. The dilapidated state of many of the remaining ships represents a growing environmental threat, and the high cost of their maintenance, plus the high cost of demolition in US yards, renders their lay-up an extremely expensive business. See Bhattacharjee (2009), p. 201.

[9] *The International Convention for the Safety of Life at Sea*, 1960.

[10] Other disposal options include sinking ships to form artificial reefs (reefing), a practice more likely carried out with ex-government-owned ships, when extensive and expensive decontamination may be carried out. Simple abandonment is another alternative, the ease with which an owner's identity may be hidden being facilitated by certain open registers—see Chap. 4.

between surplus tonnage in the freight market and the demand for scrap in the breakers' states.

A trend that has emerged in relation to ships being despatched to the breakers would appear to demonstrate the growing effectiveness of the international inspection regime of Port State controls in removing sub-standard ships from the world's fleet, such controls helping to offset the inadequacies of flag states with regard to inspection. Of the 293 ships sent to the breakers in 2006, some 98 of them (33 %) had been detained with their crews the previous year in ports around the world—and especially in Europe—for not conforming to international maritime security regulations.[11] In 2007, the proportion and number had risen to 41 % of 288,[12] 47 % of 456 in 2008,[13] rising to 58 % of 1,006 in 2009.[14] An analysis of details on the Paris MoU website is summarised below in Table 1.1 and offers a snapshot of the flags with more than one ship on the Paris Black List as at 16.2.2011, together with the main reasons for their inclusion.[15] It should be noted that although Panama has the greatest number of instances, the size of its fleet is far greater than those of the other states included.

Naval ships are often demolished by their own state although, especially in the USA, warships may be sunk as artificial reefs (after extensive decontamination) or even donated as museum ships. The decontamination of warships may be a particularly extensive exercise, given the level of hazardous materials contained within their highly subdivided structures. In December 2012, the first nuclear-powered aircraft carrier, the *USS Enterprise*, was deactivated; the whole process of decontamination and scrapping expected to take some 8 years.[16] The sinking of warships as targets as part of naval exercises has also been a common disposal route, but this practice is now being challenged in the US court by Earthjustice on behalf of the Basel Action Network and Sierra Club.[17]

A shipowner usually has two options for disposal by scrapping—either direct sale to a shipbreaker or, in some 95 % of cases, to a cash buyer, the latter providing a link between the shipowners (operators), who are looking for an immediate cash payment for their vessels, and the breakers who pay on the deferred terms of letters of credit issued by their local banks.[18] This may be particularly important for

[11] Robin des Bois (2007).

[12] Robin des Bois (2008).

[13] Robin des Bois (2009).

[14] Robin des Bois (2011).

[15] Paris MoU (2012).

[16] Ships Monthly (2013a).

[17] *Basel Action Network and Sierra Club v US Environmental Agency 2011*, filed in the US District Court of Northern California December 2011 by Earthjustice on behalf of the two plaintiff organisations. The action claims that the EPA failed to regulate adequately the dumping of PCBs claimed to be contained in former warships sunk as part of the SINKEX programme. Ships Monthly (2012).

[18] In accepting a letter of credit, the cash buyer can either wait some 180 days for his money from the issuing bank or discount its net present value at another bank.

Table 1.1 Paris MoU Black List as at 16.2.2011

Flag State	No. of ships
Sierra Leone	2
Bolivia	2
St Kitts Nevis	3
Georgia	3
North Korea	4
St Vincent and the Grenadines	5
Comoros	7
Cambodia	8
Panama	15
All other states	12
Flag not mentioned	19
Reason for banning	
No valid International Safety Management	6
Jumped detention	12
Multiple deficiencies	22
Failed to call at indicated repair yard	40

Source: Paris MoU (2012)

owners of smaller fleets, who may have only infrequent or spasmodic dealings with breakers. The use of cash buyers has further advantages to the owners of not having to transport their ships to the breakers, the vessel being sold on an 'as is, where is' basis, or of being caught up in renegotiations with breakers should trading conditions become suddenly adverse. The cash buyers run the risk of sudden and sharp declines in scrap rates in the time between purchase and resale of ships; by taking this risk the cash buyer effectively acts as an underwriter for the owner. In addition, cash buyers are now regarded in both the *HKC* and the proposed EU Regulation as full owners of their purchases with the same liabilities as previous operators, whereas previously their often short-term ownership was not so regarded. The number of back-to-back deals where the cash buyer sells immediately to a specific breaker is now declining. Not only are cash buyers providing a bridge between shipowners and breakers, but they are also entering into a greater involvement in the industry with ownership of shipbreaking yards—upstream and downstream investment.[19] A more recent addition to the disposal options has been demonstrated by the emergence of organisations, which not only arrange contracts between owners and breakers, but also play a positive role in monitoring the disposal operations and providing documentary proof of the process; such operations may be defined as 'facilitated disposals' and are examined in Chap. 7. Similarly, a small number of cash buyers will undertake surveys of breakers' yards for acceptable operators and will not sell to certain breakers; on the other hand, a shipowner may determine that his ship goes to a specific breaker, despite the cash buyer's preference.

The seller's broker, who does not take title to the vessel, is usually responsible for the drawing up of the contract. Memorandum of Agreement (MoA) between

[19] Sharma (2011).

cash buyer and shipowner instigates a move of deposit funds from the bank of the former to the latter. A subsequent MoA between cash buyer and shipbreaker triggers a letter of credit to the cash buyer's bank, which in turns releases the balance of monies due to the owner's account. Russian owners tend to prefer sales directly to shipbreakers; for this reason, sales of Russian-owned ships are not always included in shipbreaking statistics. Similarly, naval vessels tend not to be sold via the broker system.

The volume of ships offered for scrap usually follows an inverse relationship to the spot rate freight prices; when the demand for shipping services grows faster than the rate of growth of the supply of ships, then freight rates increase and the number of ships sent for demolition tends to fall, a number of ships that might have reached the end of their normal working lifespan being kept in operation whilst still seaworthy and the market rates support their use.

Conversely, when there is a surplus of ships available beyond the market demand, freight rates fall and the number of vessels for disposal may rise accordingly.

The price paid for surplus ships would appear to be a driver for the owner's decision to scrap, but price per light displacement tonnage (LDT) is negatively correlated to the volume available, freight rates—and consequently the potential earning power of the ship—still being the dominant driver. The international financial crisis of 2009, which resulted in some 700 container ships being laid up at the end of the year, saw owners despatch many of their older vessels to scrap, to maximise the benefits to be obtained from their newer (and usually more economical) tonnage. Yet despite this wave of ships for disposal, prices offered by Asian breakers rose from a low of US$200 per tonne at the beginning of 2009 to almost US$300 in December 2009.[20] Although freight rates might determine *when* a ship is dispatched for scrapping, it is the price of steel scrap that determines just *where* it is scrapped, since prices can fluctuate rapidly between places and times, due to differences in labour costs, import duties etc., as well as the local demand for steel products, usually from the local construction industries.[21]

Rendering a ship gas-free prior to demolition is an expensive operation and India's enforcement of this requirement resulted in many of the larger tankers being directed towards Bangladesh and Pakistan, where such requirements are enforced to a lesser extent. Ships are sold by light displacement tonnage, which represents the actual mass of the ship itself that the breaker buys; this is really the only time such a measure of a ship is used. The price paid by the breakers may be less than the cost of a whole year's insurance under trading conditions.[22]

[20] Robin des Bois (2010a).

[21] The massive rise in demand for steel by the Chinese building industry, plus a limited supply of vessels of all types during a period of high freight rates resulted in a tripling of scrap prices from US$125 to almost US$400 from the beginning of 2002 to the beginning of 2004. European Commission (2004), p. 70.

[22] Røst (2003).

There are, in addition, a number of other factors that can affect the price that the owner eventually receives for his surplus ship. The incidental contents that may be aboard at the time of disposal may be deemed significant to the shipbreaker, such as the high level of residual bunkers or iron ballast that may remain on board or the high specification tanks of LPG or chemical carriers. Specific vessels may also demand high prices; whilst price rates from India were around US$400 per LDT in May 2010, one specialist vessel was able to command a price of more than US$800 per LDT from an Indian breaker.[23] In 2007, stainless steel in two chemical tankers generated a price of US$1,250 per LDT from Indian breakers.[24] Additionally, a scheduled wave of new build deliveries can prompt owners to release obsolete vessels.

In 1997, a rise in price to US$600 LDT in Bangladesh led to the creation of a local cartel of shipbreakers wishing to impose a price decrease; this effectively blocked the purchase of all ships in Bangladesh until prices had fallen to under US $500 by the end of that year.[25] The strikes and legal uncertainty experienced by the shipbreaking yards of Bangladesh (see Chap. 3) effectively removed the country from the shipbreaking market, since sale and purchasing and beaching were placed on hold for an interim period.[26] A fall in rates of 16 % in 1 week offered by China, plus a steep fall in the value of the rupee against the US dollar in India, together with a forthcoming budget and the monsoon season, also impacted adversely upon the market.[27] In such circumstances owners may have to face a renegotiation of prices, especially of high prices, that have previously been agreed in what are known as ‘beachfront negotiations.’ Such situations can also affect cash buyers, who may suddenly find that they are unable to offload new purchases at previously agreed prices.[28] Cash buyers (and breakers) can also fall foul of the issue of last minute arrest warrants for ships whose owners have large financial liabilities, effectively tying up their cash and assets possibly for extended periods.

The years 2004–2007 were boom years for the shipping industry, with a high demand for ships for trade, older ships being kept in trading and relatively few going for scrap, banks with little or no shipping experience setting up shipping operations and cash buyers becoming owners. The collapse of the shipping market in 2008 saw the re-negotiation of many charter contracts and banks began calling in loans and withdrawing the supply of finance; a range of financial institutions active in 2008 are no longer around today.[29] On balance, it is adverse economics that renders many ships redundant, rather than ships coming to the physical end of their lives.[30]

---

[23] Reyes (2010a), p. 12.

[24] Robin des Bois (2008).

[25] Robin des Bois (2008).

[26] Robin des Bois (2008).

[27] Robin des Bois (2008).

[28] Reyes (2010b), p. 9.

[29] Wansell (2011).

[30] Comment by Ansari (2012).

A further factor that transcends purely economic considerations in the flow of ships to the breakers is the impact of specific legislation such as the separate decisions by the USA, Europe and the IMO to limit the life of single-hulled tankers—see Chap. 3—and see them replaced by double-hulled tankers; however, by the end of 2011, double-hulled tankers themselves were already being sent to the breakers. Legislative restrictions do not necessarily need to be directly impacting, but can arise from more indirect measures such as the establishment of emission control areas or changes to the allowed levels of sulphur content in fuel oils.

The factors surrounding the decision to dispose—and to purchase—may therefore be summarised on the following chart, Fig. 1.1. Combining these two relationships—it is the freight rate that determines the number of ships available for scrap, whilst it is the local demand for the scrap produced from the ships that often sets the price of scrap, moderated by the level of competition between the breakers. A sudden increase in the price of scrap or a fall in the demand for steel products could, however, result in the breakers suddenly submitting their purchases to a close scrutiny against the details of the contract in the hope of renegotiating a more favourable price.

There are no prescribed international standards relating to required documentation or conditions when a ship is sold for scrap. As a result, the shipowner traditionally has had little or no incentive to either undertake any pre-cleaning or document the accumulated hazardous materials that may be on board.[31] The first, but not particularly successful, standard contract for vessels for demolition was the SALESCRAP 87, which was prompted by the increase in scrapping in the 1980s.[32] Since 2001, the standard contract for the sale of a ship for scrap has been BIMCO's DEMOLISHCON,[33] with the seller being only obliged under 'best endeavours' to give such information as recommended by the *IMO Guidelines on Ship Recycling*,[34] and the buyers to use their best endeavours to comply with them. For this reason, the use of the DEMOLISHCON contract is not high, although its use was endorsed by the IMO in 2003. Foreseeing the effective introduction of the *HKC*, the proposed RECYCLECON, adopted by BIMCO in 2011, represents a contract format specifically aimed at 'green' recycling, which is currently a niche market but expected to grow with the Convention. The new format aims to improve standards overall by reflecting 'green' practices and making liabilities and obligations as clear as possible through unambiguous wording.[35] The formal introduction of mandatory provisions of the *HKC* could lead to higher contractual conditions of sale, including an inventory of hazardous materials; until then, cash buyers and breakers will continue to compete for surplus ships on the basis of price alone.[36]

[31] Andersen (2001), p. 19. This need for documentation will change once the *HKC* enters into force.

[32] Hunter (2011).

[33] This contract format is adaptable for use in either an 'as is, where is' or 'delivered basis' sale.

[34] International Maritime Organization (2003).

[35] Hunter (2011).

[36] Drury (2011).

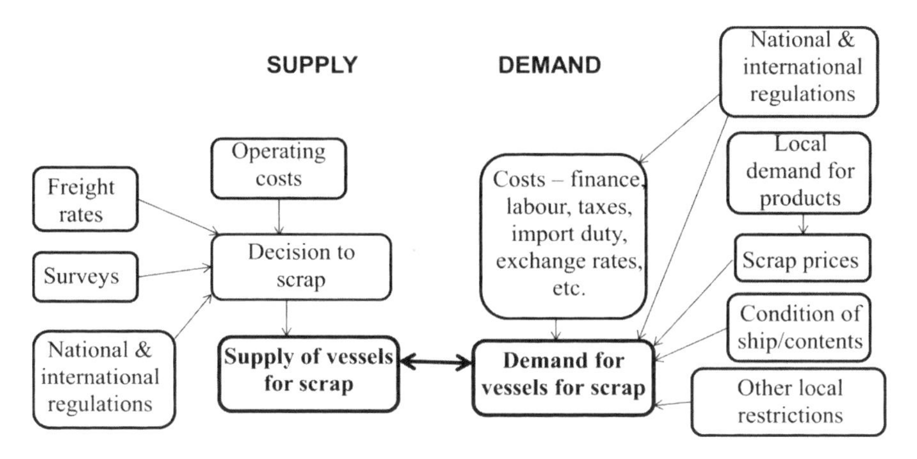

Fig. 1.1 Key drivers for the supply and demand of vessels to the shipbreaking industry. Source: Galley (2013)

Inspections of vessels sold for demolition are usually of a much reduced level than if the ship were sold for further trading.

Once a ship has been purchased, the cash buyer may then crew and insure the vessel, often reflagging and renaming it, and delivering it to the successful shipbreaker. A ship in this process may be resold, reflagged and renamed several times before it reaches the scrap site, at times in a deliberate attempt to conceal the identity of the owner or the ship itself—see Chap. 4. A transfer of ownership, however, involves the transfer of liabilities for a ship and its contents; hence owners may wish to sell a surplus ship just prior to the end of its working life in order to pass on any liabilities that may arise for pre-cleaning the ship of hazardous content, before its despatch to the breakers. There are also numerous instances of ships arriving at scrap sites with false documentation as to ownership, or registration, and even under the flag of a fictitious state.[37] Since the breaker is operating against letters of credit issued by a bank, he is under pressure to demolish the vessel and realise his return as quickly as possible, in order to minimise the level of interest due to the bank; it is therefore important for a cash buyer to have a breaker ready.

Figure 1.2 below illustrates some of the major influences and pressures which act on shipbreakers.

The appropriate clearances having been obtained, a ship may be allowed to enter territorial waters for beaching. Depending on the state of the tides, the ship may lie at anchor offshore until the appropriate high tide occurs. In this event, demolition cannot begin immediately, yet interest is accruing; the timing of the completion of transactions is therefore an important consideration to the breaker. The state of the

[37] See Chap. 5. The *Platinum II* arrived for demolition at Alang in 2009, falsely registered under the flag of Kiribati, whilst in 2006, the former Danish ferry *Riky* reportedly arrived under the flag of the fictitious state of Roxa.

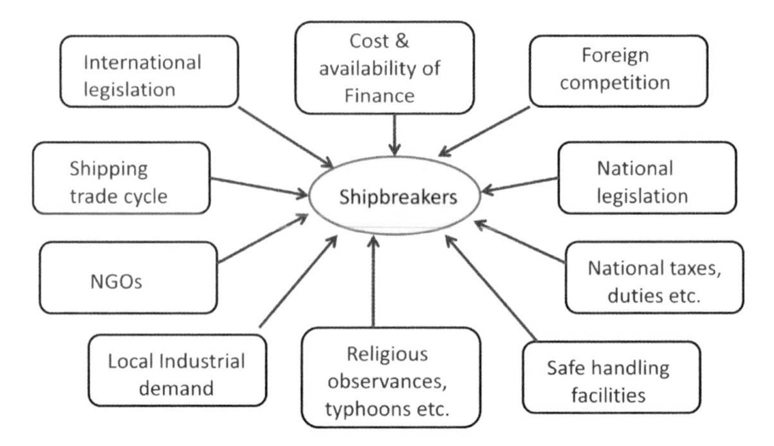

Fig. 1.2 Summary of pressures and influences on shipbreakers. Source: Galley (2013)

international economy can have a further impact upon cash buyers and shipbreakers alike, and the question of financially distressed owners can prove to be a minefield for shipbreakers. Agrawal, legal advisor to a leading cash buyer, considers that the failure of owners to disclose the sale of ships for disposal until a late stage prompts creditors to seek the issue of court orders for the arrest of vessels against which debts are outstanding after they have been sold to cash buyers or even after demolition has begun.[38] Whilst the viewpoint of the High Court in Pakistan is that the 'characteristics' of a vessel have changed by this stage and arrest is no longer an option, the Indian High Court accepts that an arrest can be enforced until demolition is totally completed, a situation that may leave either the cash buyer or breaker with a frozen asset and possible bankruptcy as interest continues to accrue.[39] There is, therefore, a need for due diligence, with a pre-contract search for debts etc. via databases such as the European Maritime Safety Agency (EMSA) Equasis database. Ships bought in judicial sales have a clean title when purchased through a court.

1.2.2 Demolition

It is the need to drive ships up onto the beach that requires that they arrive in an operational state, and this in turn may preclude the removal of much of the hazardous content, especially within the realms of the engine room. Ships that arrive under tow are much more difficult to beach, and towing itself is a slow and costly operation. In addition, there is a distinct lack of ocean-going tugs sufficient to

[38] Agrawal (2011), p. 6.

[39] Agrawal (2011), p. 6.

facilitate the necessary towage operations to deliver a large number of 'dead' tows that might arise after extensive pre-cleaning of ships prior to departure.[40] As a result, ship breakers and their agents prefer to buy ships that are anchored nearby—often lying off Singapore—or are on their final voyage, which makes the pollution-free certification laws easier to circumvent.[41] Prior to beaching, tanks may be discharged and valuable, saleable or restricted items, e.g. uncontaminated oil, electronic equipment, removed. Any accumulated waste may be subject to national law if the nation is a signatory to the relevant annexes of *MARPOL*.[42]

Vessels are beached by driving them at full power towards a narrow expanse of beach until they are firmly aground. Positioning must be exact and the vessel driven as high up the beach as possible; the time (and hence the cost) needed for demolition can be doubled if the beaching is not carried out correctly. Timing is also important, since the number of days when tides are sufficiently high as to allow beaching are probably limited to just two or three per month. Prior to the actual beaching, a scrapping plan is prepared so that the ship will remain upright and in correct balance to allow the breaking to proceed. Cutting is done by gas torches, so the vessel must be correctly gas-freed. Since this is an expensive process, it is not always done in a thorough manner and pockets of gas may remain for a large part of the process—hence the numerous instances of fires and explosions that may occur.[43]

Smaller items of gear, *e.g.* anchors and chains, are removed at an early stage; larger items such as engines are removed as they become accessible. A very high proportion of fittings and loose items are sold for re-use, often, for instance, in the numerous shops that line the road from Alang to Bhavnagar for several kilometres. This reuse is an aspect of shipbreaking in Asia that does not exist to anywhere near the same extent in other countries, where the reuse of equipment is often not considered, or the equipment requires mandatory testing. Ships are made of various grades of steel, yet their design and build often make it difficult to extract the higher (and hence more valuable) grades of metal. Projects are in hand to identify the various materials in order to maximise the returns from scrapping.[44]

[40] Opinions of IMO personnel expressed during discussions at the IMO on 1 December 2009.

[41] Hossain and Islam (2006), p. 7.

[42] International convention for the prevention of pollution from ships 1973 as modified by the 1978 Protocol MARPOL 73/78. Andersen (2001), p. 20.

[43] As a result of the deaths of six workers following an explosion on board the tanker *Union Brave* that was being demolished at the Kiran Ship Breaking yard at Alang, three yard owners were arrested on suspicion of culpable homicide and negligent conduct, a charge that carries a maximum sentence of life imprisonment on conviction. In response to the arrests, the shipbreaking yards at Alang were closed by their owners in protest. Reyes (2012).

[44] Following the example of the European car industry, metals may be identified as to different types and grades. A pilot project to test this process was scheduled to begin with the new Maersk Triple E ships in September 2012. The long term goal is to be able to take steel plate from ships under demolition and use it directly in the construction of new ships. Kimmins (2012).

Openings are made in the hull for the access (and escape) of the workers, and to allow light to enter and fumes to escape, but the very process of venting fumes may trigger the fires and explosions that the activity is meant to obviate. Starting from the bows, large sections of the ship are cut away in a manner such that they fall hull side down in order to facilitate their towing towards the beach. There they are cut into smaller, more manageable pieces by torch or by hammer and chisel and loaded by hand to trucks for dispatch to the re-rolling mills. The valuable bronze propellers are laboriously sawn apart by hand. During this process, the balance of the ship must be maintained, a situation that is somewhat more difficult to maintain if the ship is being demolished whilst still afloat.

As the ship is lightened, it is towed progressively further inshore, often using untested winches and chains that have been removed from previous vessels. Liquids such as oils and bilge water may be discharged to the beaches and to the sea throughout the breaking process. Many wastes may be burned on the beach, including electrical cabling, which is burned to remove the covering from the copper core, releasing toxic furans and dioxides in the process. Large amounts of asbestos have traditionally been removed by hand, mainly from the engine room, either for resale[45] or dumping in the sea. Protective equipment for the workers has usually been minimal or non-existent. Piles of unwanted waste often litter the surrounding countryside.

1.3 Historical Locations

Shipbreaking is a mobile, migrant industry. The UK, USA, Germany and Italy were once major centres for shipbreaking, where the industry was regarded as highly mechanised. At one time, 50 % of ships were broken in the UK, whilst Scotland was regarded as the home of the largest shipbreaking operation in the world.[46] As regulatory controls over health and safety and environmental pollution began to expand and to stiffen in these areas, operating costs began to rise as a consequence. The industry therefore sought to move to countries with lower labour costs, a high demand for scrap metal coupled with a dearth of local natural resources and a low regulatory regime. Consequently, in the 1970s, ships were being broken up in the docks of Spain, Mexico and South Korea.[47]

Associated with this transfer of the industry to the developing countries was a move away from dismantling ships alongside or in dry docks, where pollutants could be managed more effectively, and cranes could more easily handle the large sections of ship that were removed. Instead of hard standings or quaysides, beaches

[45] The use of asbestos is not banned in India and there is a ready market for the asbestos waste generated during the demolition process.

[46] Hossain and Islam (2006), p. 2.

[47] Greenpeace (1999a), p. 6.

became the norm, and the use of heavy equipment was replaced by manual labour and the use of hammers and chisels and saws.

The reduced demand for new ships after the shipbuilding boom of the 1970s prompted Japan to employ idle shipbuilding docks for shipbreaking in a 'scrap and build' programme, which it was hoped would stimulate the demand for new vessels, but this eventually proved to be uneconomical and was discontinued,[48] although shipbreaking within an enclosed dock was much more conducive to controlling the spread of pollution than breaking on open beaches. By the 1990s, Far Eastern countries had become predominant in the scrapping of ships.

Today, shipbreaking is concentrated on the beaches of the Indian sub-continent, principally at Alang (India) and Chittagong (Bangladesh), with Gaddani Beach (Pakistan) playing a somewhat more intermittent role. Here, during the short periods of highest tides, ships are driven up onto the low gradient, intertidal zones, where they are demolished, the ships being progressively winched further up the beach as demolition progresses. Scrapping is largely done by hand and the discharge of pollutants into the sea and the atmosphere can be extensive. Since 2004, more than 80 % of the larger end-of-life ships have been dismantled in India, Bangladesh and Pakistan. China has also emerged as a major site for the industry, but breaking is carried out mainly along river banks, where the process can be somewhat more controlled. A relatively small, but nevertheless important, industry also operates at Aliağa, on the coast of Turkey. All these centres operate in competition with each other and their individual dominance depends upon a variety of factors such as the demand for rebar from the local building industry and national taxes and customs levies effective at any particular time. Statistics relating to numbers of plots, workers employed etc., vary over time and between reports. As a result of selling ships to these yards, the shipowners receive a much higher sum for their redundant vessels than if they were sold to a Western nation for scrapping, although the greater controls over the Chinese and Turkish yards result in a somewhat lower price being offered than is the case with India, Bangladesh and Pakistan. The size and depth of any European sites which still practice shipbreaking are also quite inadequate for the size of some of the vessels that now have to be accommodated.

A simple transfer of technology from the more industrialised nations to the shipbreakers of the sub-continent, would not necessarily lead to a major advancement in the way ships were broken there if the context remains unaltered. To be employed successfully, state-of-the-art western technologies are also likely to require western legislation and enforcement and a trained and aware workforce.[49]

---

[48] Similar 'scrap and build' programmes were introduced in the 1930s by both Japan and the UK, and recently by China.

[49] Greenpeace (2002a).

Table 1.2 Recycling locations in terms of percentage of total recycling 2012

Location	Method	%	No. ships
India	Beaching	40	497
Bangladesh	Beaching	18	230
Pakistan	Beaching	10	124
Sub-total		**68**	851
China	Afloat	17	209
Turkey	Landing	12	153
EU		2	28
Rest of World		1	13
World Total		**1,254**	**100**

Source: Adapted from NGO Shipbreaking Platform (2013)

1.4 Current Major Operations

This section describes the current major shipbreaking locations—mainly in South Asia—and the manner in which the demolition activities are carried out. It concludes with a short description of some other traditional and potential sites. The relative importance of South Asia is aptly shown in Table 1.2.

1.4.1 India

India's principal shipbreaking site is located on the western coast of the Gulf of Khambhat (Cambay), Gujarat. The site chosen was the result of specific policy decisions taken by the Central and Gujarat State governments desirous of diversifying the industry away from its then established site at Mumbai (Bombay), following the adoption by the Central Government of a 'shipbreaking development fund' in 1978 to facilitate the importation of ships for scrapping by the Metal Scrap Trade Corporation.[50] Some 56 km south of Bhavnagar, the site has the natural characteristics that make it an ideal place for the beaching of large ships—a long sloping continental shelf, a high tidal range of up to 13 m, and relatively mud-free conditions, which together allow ships to be beached during the high tides,[51] but leaving exposed a wide inter-tidal zone which facilitates operations between the tides. As demolition of the vessel progresses, ships can be slowly winched further up the beach, the wide waterfront serving as a scrap collection area for the further dismantling of large sections of the ships as they are removed from the hull. The activity actually occurs on two adjacent sections of the beach at the Alang and Sosiya Shipbreaking Yard (ASSBY), covering some 15 km in length; hereafter, the site will simply be referred to as Alang.

[50] UNESCO (2004), Chap. 1.

[51] UNESCO (2004), Chap. 1.

From the arrival of the first vessel for demolition at Alang, the beaches have grown to constitute the largest shipbreaking region in the world. Since 1992, Alang has dismantled more than 5,000 ships, representing some 35.6 million LDT, with a peak of 361 ships in 1999.[52] Shipbreaking at Alang was recognised as an industry by the Government of India in 1979. Each nation's shipbreakers tend to have their own preference for the types of ships they handle; the Indian yards tend to prefer specialised ships, such as passenger ships, ro–ros and reefers.[53]

Prior to the establishment of shipbreaking there, Alang and an area of some 12 km around it comprised of some ten villages, whose inhabitants followed a life of subsistence farming. The 1961 population, estimated at 7,000–8,000, had grown to almost 20,000 by 1991[54]; today, the size of the population has doubled again and figures of 30,000–40,000 are often cited. When the industry was at a peak in 1998–1999, some 60,000 shipbreakers were employed.[55]

The whole of the coastal zone operations lies under the management of the Gujarat Maritime Board (GMB),[56] which appears to have had major shortcomings in meeting its responsibilities for work safety, as required by the *Factories Act 1948* (as amended 1987) and the standards for the safety of both workers and waste disposal.[57] Little or no compensation has traditionally been paid to injured workers. The United Nations Educational, Scientific and Cultural Organisation (UNESCO) survey of 2004 reported a distinct lack of spatial planning with regard to housing, roads, drainage and sewage disposal, water and electricity as well as an absence of infrastructure provisions for hospitals, schools etc., stating that the GMB has neither the capacity nor the training for the provision of infrastructure facilities.[58] Regulations laid down to address problems of safety and environment have had little implementation success in the past, with allegations that the powers enjoyed by the GMB were limited by '*the economic and political clout enjoyed by the ship-breakers.*'[59] These were the privations experienced by the population of the area, even prior to the specific hazards and pollution that shipbreaking brings. A report issued by the Central Technical Team of the Ministry of Environment and Forests in 2009 described the nature of shipbreaking at Alang being such as to render it hazardous due to the heights at which workers had to operate; their lack of training; the presence of hazardous materials and the risks of fire and explosions with which they had to contend.[60]

---

[52] Vora (2010).

[53] Roll-on, roll-off vehicle ferries and refrigerated cargo ships (although the latter are now a declining type, due mainly to the use of refrigerated containers).

[54] UNESCO (2004).

[55] A report by the International' Federation, 2006 cites a total of 55,000 workers at Alang, plus 6,000 at Mumbai, with a further 160,000 employed in downstream activities. IMF (2007).

[56] With the exception of the port of Kandla.

[57] Dube n.d.

[58] UNESCO (2004), Chap. 1.

[59] Abdi (2003).

[60] Central Technical Team (India) (2009).

In 2001, some 173 plots were available at ASSBY, although the number actually in operation at any one time may vary, dependent upon the vicissitudes of the industry. In 2006, the number of active plots had fallen to just 25, by 2009 it had risen back to 130. In 2010, however, only 20 yards were reported to be in operation again.[61] Some 10 plots were dedicated for the demolition of Very Large- and Ultra Large Crude Carriers (VLCC and ULCC), although the majority of these ships are now demolished elsewhere.[62] Plots are leased from the State, initially for 5 years, with a provision for renewal for a further 10 years.[63] Allocation of plots is carried out under the provisions of the 2006 *Ship Recycling Regulations*[64] enacted under the 1981 *Gujarat Maritime Board Act*. Indian yards are prohibited from working on more than one vessel at a time.

Although the biggest shipbreaking site, and almost synonymous with the shipbreaking industry, Alang is now secondary to Bangladesh in terms of annual tonnage scrapped.[65] Alang is also in danger of losing its premier position to China, where the government is acting in a more positive manner in terms of support for its shipbreakers, and where, despite the pollution also engendered there, the country is becoming more appealing to shipowners wishing to establish longer term arrangements with specific yards. Consequently, the Gujarat Maritime Board was reported to have begun a programme of infrastructure developments and treatment, storage and disposal facilities.[66]

The labour force is overwhelmingly composed of migrant workers, mainly from the poorer states of Bihar, Orissa and Uttar Pradesh, predominantly young men, with a high rate of illiteracy, and employed on a contract basis, although there is no written contract. Few women are employed in this industry and in India persons below the age of 19 are not allowed to take part.[67] The wages paid to this labour force are low by most standards, but certainly represent an improvement on earnings potential in their native states.[68] There is a distinct lack of representation amongst the workers, trades unions being effectively banned.[69] The level of disease

---

[61] Pandey (2010).

[62] India's requirement that tankers arrive gas-free ultimately led to many of these vessels being diverted to Bangladesh, where such regulations were not enforced with any regularity. The cost to the owners of certifying a gas-free ship is a cost that is not incurred with vessels heading for other sites on the sub-continent.

[63] Gujarat Maritime Board (2007).

[64] 2006 *Gujarat Maritime Board (Conditions and Procedures for Granting Permission for Utilising Ship Recycling Plots) Regulations*.

[65] This was prior to the closure of the yards at Chittagong by the Bangladesh High Court in 2010–2011.

[66] Basu n.d.

[67] Putherchenril (2010), p. 35.

[68] In addition to those employed directly in the scrapping process, there are a large number of other workers, mainly from Gujarat, working in ancillary activities. UNESCO (2004), Chap. 1.

[69] This is in contravention of article 19 of the *Constitution of India*, the *1926 Trades Union Act*, and the *1947 Industrial Disputes Act* etc.

resulting from the pollution affecting both the crowded and unsanitary settlements and the beaches themselves is high. A committee set up by the Supreme Court in 2006 commissioned the National Institute of Occupational Health (NIOH) to examine working conditions at Alang and concluded that one in six of the workers examined showed signs of asbestosis, a whole of other diseases arising out of the unsatisfactory conditions that prevailed.[70] The Alang Ship Breaking Association denied the findings of the report.[71]

Injuries and death arise from a wide range of incidents, including fire and explosions, falling from heights inside the ships or to the ground, falling objects, burning and eye injuries from cutting operations oxygen deficiency in confined spaces, snapping of ropes and chains, and lack of personal protection equipment (PPE) etc. The above report submitted by the NIOH to the Supreme Court also noted that the fatal accident rate in shipbreaking at Alang stood at 2 per thousand, compared with a rate of 0.34 per thousand in the mining industry. The prevalence of workers willing to carry out demolition meant that those so engaged were often regarded as '*replaceable commodities*.'[72] The lack of medical facilities, training, safety equipment and personal protection equipment have all been repeatedly criticised, but the consensus is now that these have all started to show improvements in recent years.

'A report from the Wise Coast Practices for Sustainable Human Development Forum in India (1999–2000) called for the application of the then-in-force *Interstate Migrant Workman Act* to address the living and working conditions under which the workers survive.[73] A later IMF report on shipbreaking labour in 2007 defined shipbreaking as one of the most hazardous occupations within the hazardous waste section. Since the industry was not at that time classified as an industry, any provisions and benefits offered by the *Indian Factory Act* of 1984 were not applicable to those employed in shipbreaking.'[74]

In 1998, Greenpeace visited the shipbreaking yards at Alang and Mumbai. The highly critical report[75] that emerged, re-enforced by a second report[76] from a subsequent visit made in 2002 at the invitation of the Gujarat Maritime Board, led to the GMB designating the shipbreaking sites as Restricted Areas, with access only by permit from the GMB, and subject to various prohibitions. Continued criticism from numerous media reports resulted in a ban imposed in 2010 on all journalists from visiting the sites.[77] However, not all Indian sites operate at the

[70] Abdi (2003).
[71] Infochange India (2011).
[72] Putherchenril (2010), p. 34.
[73] Joshi (1999).
[74] IMF (2007), p. 3.
[75] Greenpeace (1999a).
[76] Greenpeace (2003).
[77] Shashikumar (2005).

same level, and there are now some yards where well paid and well trained staff employs safe, modern methods.[78]

The procedure for allowing ships to be broken up at sites in India stems from Writ Petition No. 657 of 1995, *Research Foundation for Science Technology National Resource Policy v. Union of India & Ors.*, the Supreme Court of India issuing an order on the subject of the dumping of hazardous wastes on 14 October 2003. On the recommendations of a specially formulated High Powered Committee (HPC), the court ordered that ships wishing to enter port (or beach) for scrapping should have the proper consent from the appropriate authority certifying that no hazardous or radioactive materials remain on board.[79]

The Atomic Energy Regulatory Board (AERB) is to be consulted where appropriate. Further, ships should be '*properly decontaminated by the ship owner, prior to breaking*,' in these instances, the term 'shipowner' may be taken to mean the last owner of the ship that is about to be broken, *i.e.* the shipbreaking operator. Responsibility for ensuring that this decontamination is carried out lies with the State Pollution Control Board (SPCB).[80] The ship's owner is also responsible for providing a complete inventory of hazardous waste on board, which is a prior requirement for the granting of breaking permission.[81] A summary of the authorization process at Alang once a letter of credit has been opened by the breaker is as follows:

- Documents are submitted to Customs, the State Maritime Board (SMB) and the State Pollution Control Board (SPCB), who carry out a desk review
- The ship arrives at an outer anchorage before being allocated a berthing position by the SMB
- Inspections are undertaken by Customs, the SPCB and by the Atomic Energy Regulatory Board (AERB)
- Once all Customs dues have been cleared, the ship is formally delivered to the breaker and beaching may proceed
- The vessel must be cleaned for the issue of confined space and hot work permits and the SPCB inspects for decontamination
- A ship recycling plan and other documentation is submitted to the SMB for final permission for breaking to begin.[82]

Other provisions relate to the authorisation of shipbreaking sites and the handling and disposal of hazardous wastes. Only sites with '*provisions for disposal of the waste in environmentally sound manner*' should be authorised to operate, under Rule 5 of the Hazardous Waste Rules, 2003.[83] Prior to the opening of an asbestos

[78] McCarthy (2010a), p. 4.

[79] Supreme Court of India Order 14 October 2003 s.70.2(2) 1.

[80] Supreme Court of India Order 14 October 2003 s.70.2(2) 2.

[81] Supreme Court of India Order 14 October 2003 s.70.2(2) 13.

[82] Jani (2011).

[83] Supreme Court of India Order 14.10.2003 s.70.2(2) 5.

segregation centre, a report[84] from the NGO Young Power in Social Action (YPSA) described the waste disposal system of the GMB as a dump disposal system whereby, from 2006, waste was dumped in large holes lined with plastic, which were subsequently covered with soil once full; another hole was then dug to continue the practice. Currently, much of the disposal work with hazardous waste is now contracted to the Gujarat Enviro Protection and Infrastructure Ltd. (GEPIL). Unfortunately, during the pre-beaching inspection of ships, hazardous wastes are only recognised (at least for purposes of securing beaching permission) if they are deemed to be part of the *cargo*; hazardous materials built into the structure of a ship-for-disposal are not generally considered wastes. There are also other instances where, if the situation of a ship for disposal does not meet the appropriate regulations with regard to its entry, then the answer appears to have been to amend the regulation, rather than address the problems of the ship. These aspects of regulation are considered further in the case of the *Platinum II*, in Chap. 5.

Following a major explosion on a tanker under demolition in 1997 in which many were killed,[85] gas-free requirements for ships for demolition were imposed in India. The Notification issued by the Gujarat Maritime Board in 2001 on Gas Free for Hot Work certification is reportedly now enforced, and its observance has played a major role in shifting the scrapping of certain ships—especially tankers—to Bangladesh, where the enforcement of such requirements was low.

Although numerous reports speak of improvements introduced in training, the use of PPE, and attempts by certain yards to address the question of environmentally sound shipbreaking, there are still reports of the extremely poor conditions at Alang. As late as January 2010, the Special Rapporteur of the United Nations Human Rights Council, Okechukwu Ibeanu, reported that although he recognised the progress made in the management and disposal of hazardous products and wastes and that India had:

> ...developed a comprehensive legal framework to protect human rights and ensure the environmentally sound management of hazardous products and wastes,

he continued by saying that with regard to shipbreaking:

> National legislation on waste management and health and safety at work is not effectively implemented, and the current institutionalised framework appears inadequate to respond to the health and environmental challenges posed...The health and safety situation prevailing at the shipyards continues to remain critical, especially in Mumbai where the working conditions and the quality of facilities remain highly inadequate for guaranteeing health and safety at work and an adequate standard of living for those employed in the shipbreaking sector.

He was also:

[84] Young Power in Social Action (2007). YPSA is a member of the NGO Shipbreaking Platform.

[85] 16 deaths according to the authorities, 30 deaths according to the workers. Bari (2006).

> ..."shocked" by the extremely poor conditions in which most workers live in Alang and Mumbai ...[where workers]... live in makeshift facilities lacking basic sanitation facilities, electricity and even safe drinking water.

He called upon the Government authorities to provide appropriate plots of land and to facilitate the construction of adequate housing as well as adequate sanitation and drinking water.[86] This report was not supported at the IMO.[87]

Whilst conditions at Alang may still leave room for improvement in many areas, there is growing anecdotal evidence to say that some improvements have been made in recent years, with provision of more mechanical handling equipment and better facilities in the medical and fire services. There is also a requirement for basic training for the workers before they are allowed to take part in the breaking operation, two certificates being necessary to demonstrate instruction in health and safety and in the operator's own trade (specific groups of people work on specific aspects of the ships).

Large though it may be, Alang is not the only site in India where shipbreaking activities are carried out. Before the emergence of Alang, India's main shipbreaking activities were locate at Mumbai (Bombay) in Maharashtra, along a waterfront area of Darukhana, north east of the city, where shipbreaking is still undertaken on land belonging to the Mumbai Port Trust (MPT) and under the control of the Ministry of Shipping. The industry began operations in 1910 and, as at Alang, ships are driven up onto the beach at high tide to be cut up at low tide. Again, these sites are now also designated as Restricted Areas.

In a survey[88] commissioned by the IMF into a comparison between workers' conditions at Alang and those at Mumbai in 2001, it was concluded that for the 15,000–16,000 workers at the (then) 16 Mumbai yards, conditions of housing, sanitation, food and nutrition, and health and safety facilities were all cited as even below the standards found at Alang (itself deemed no model of good standing by writers at the time—and subsequently). Working hours at Mumbai were deemed longer and wage rates and minimum wages were also below those at Alang, and distinctly below those quoted by the owners of the Mumbai sites.

Possible reasons offered for the disparity between conditions at the two locations are spread between the regulators and the regulated. The Gujarat Maritime Board, with responsibility for operations at Alang, was depicted as somewhat more go-ahead than the MPT and more instrumental in advancing the business, both in terms of infrastructure and the extent of the industry. Activities at Mumbai are on a much smaller scale than at Alang and the physical potential for expansion is limited by the yards' proximity to the city and the consequential restrictions on land both

[86] Ibeanu (2010), p. 14. The report covered a visit to India to inspect the impact of both the shipbreaking and electrical and electronic wastes industries.

[87] Discussion with an IMO representative, 15 June 2011.

[88] I-Maritime Consultancy (2001).

for housing and for expansion. The result is '*congested dwellings, dirty roads, lack of infrastructure*.'[89] The much shorter rental period for the Mumbai plots was quoted, at that time, as being 3–4 months, a period that would probably discourage most efforts at relieving conditions.

A subsequent survey undertaken jointly by the International Metalworkers Federation (IMF) and Federati Nederlandse Vakbeveging (FNV)[90] in 2006, confirmed that conditions at Mumbai had remained unchanged. The survey found some 6,000 workers engaged in shipbreaking at Mumbai, again consisting mainly of young and illiterate migrant workers from the poorer states of India. Workers in the downstream industries of local rolling mills, foundries, oxygen plants and other smaller businesses brought the total figure to around 20,000. The slum-level housing conditions of the workers, with the high levels of congestion, dirt and stench, and a very low level of sanitation, was attributed to a physical lack of space and a high level of government agencies' apathy, the latter leaving the site operators to maximise profits via a sub-contracting of work system, and placing the burdens on the workers themselves.[91] As at Alang, workers are mostly casual, hired on a daily or monthly basis, or on a contract for a specific vessel; they have no written contracts or security of employment, nor do they have direct access to the employer, since mediation is through the contractor (*muqadam*). Hours are long, often varied, and usually contain two hours of compulsory overtime.

In a report by Greenpeace, who visited Indian breakers in 1999, it was stated that at Alang, blue asbestos was removed by hand, without any protective equipment, and laid out to dry for later use,[92] whilst at Mumbai, women carried asbestos containing materials to dump in the sea Fuel remnants and bilge oil were also pumped directly into the sea at Mumbai, and parts of the beach had disappeared under a three foot high layer of waste.[93]

A follow-up visit by Greenpeace in 2002 found the risks from toxic wastes and the absence of training on how to deal with them, unsafe working conditions, and lack of personal protection equipment, still unchanged. The shipbreakers who rented the plots paid no heed to the safety of their workers and the MPT, responsible for supervising the rules for shipbreaking, was criticised for '*failing to implement basic safety measures and environmentally sound shipbreaking practices*.'[94] It was further reported that the MPT gave financial incentives to shipbreakers to break up a ship quickly, and the Greenpeace conclusion was that such an incentive based on

[89] I-Maritime Consultancy (2001).

[90] International Metalworkers Federation (2007).

[91] International Metalworkers Federation (2007), p. 6.

[92] In one ship being demolished at Alang, it was observed that dusty blue asbestos was so prevalent that it was collected in open bags and sold in a nearby shop. Greenpeace (1999a), p. 11.

[93] Greenpeace (1999a), p. 25.

[94] Greenpeace (2003), p. 3.

speed, rather than safety, could only add to the avoidance of health, safety and environmental standards.[95]

In 2009, the MPT sought to rationalise shipbreaking in the area and ameliorate the problems of pollution that the industry generated[96] Accordingly, shipbreaking activity was to be allowed at a maximum of only six sites, at plots at Lakri Bunder South. A maximum of two extra ships were to be considered for beaching on a first come, first serve basis, whilst six ships were being demolished, on condition that work on one started as soon as the breaking of another had been completed.

A little to the south of Alang lies the port of Pipavav. Pipavav is distinctive in that it contains modern facilities for the demolition of ships, especially big, single-hulled tankers, under controlled conditions, with sheds, sewage treatment plant and oil separators available. The ships would be demolished whilst in dock, but still afloat, limiting the release of pollution directly into the sea. The site was built in 1997 with US$90 million of Japanese financial support, but with the intention of shipping the resultant steel scrap back to Japan, whilst leaving the wastes in India, where the cost of labour is significantly below that in Japan. To date, Pipavav has received no ships for demolition, ship building and repair being found more profitable than demolition, in which the facilities cannot match the low costs of beaching.[97] It is also alleged that as a result of the failure of the port to meet its expected operations, Japan has asked for the return of its investment.[98]

In February 2012, the government of Japan signed a Memorandum of Agreement with the Gujarat Maritime Board for the funding of the development of shipbreaking at Alang—see Chap. 7.

1.4.2 Bangladesh

Shipbreaking in Bangladesh had a casual beginning when a vessel was stranded off Sitakund, just north of Chittagong, after a severe cyclone in 1965 and broken up some years later by Chittagong Steel House Company. Ships damaged during the 1971 Liberation War were also brought ashore and scrapped in 1974 by Karnafully Metal Works. During the 1980s, the industry flourished, with nearly 70 operators working; business was booming and the alleged corrupt practices with regard to the use of government loans led to a tightening up of regulations, with the imposition of a government tax on the industry and a more conservative approach to money lending by the banks.[99] In 2009, Bangladesh shipbreaking yards accounted for one third of the world's shipbreaking tonnage, with some 20 yards, or plots, in

[95] Greenpeace (2003), p. 5.

[96] Mumbai Port Trust Docks Department (2009).

[97] Uytendaal (2012). For ISRA (International Ship Recycling Association), see Chap. 7.

[98] Comments made during discussion with government advisor 17 January 2011.

[99] Hossain and Islam (2006), p. 6.

operation. Until 2011, shipbreaking had not been recognised as an industry in Bangladesh, hence it had received no government support in the form of grants, tax relief or infrastructure development. More mechanisation is still required, as is the provision of basic facilities such as clean drinking water, fire fighting facilities etc.[100] The industry supplies approximately 80 % of the country's need for metal scrap,[101] feeding some 200 small rolling mills.[102] Unlike as in India, the Bangladeshi rolling mills do not operate a reheating process, but roll the metal cold. Rebar for building produced from scrap is significantly cheaper than products produced from steel billet[103] and both the Indian and Bangladeshi shipbreaking sites are important suppliers to their respective building industries.

Today, the industry is still concentrated in Sitakund. Bangladeshi shipbreaking yards now specialise in the demolition of the larger VLCC and ULCC tankers, which are generally avoided by Indian yards since the enforcement of gas-free requirements there.[104] The larger sites can handle two to three ships simultaneously and the yards employ some 30,000 workers directly and between 150,000 and 200,000 indirectly.

Although there are no official figures relating to the extent of child labour within the yards, a report issued by the FIDH and the YPSA in 2008 estimated the figure to be up to 25 % of the workers.[105] The influence of the plot owners over the authorities would appear to be high and it is offered that the industry needs to be bound by distinct rules rather than policies. Rather than direct action against polluters, the government has proposed 'zoning' for shipbreaking, which it defines as an emerging growth industry.[106] A failure to enforce the requirement that ships be gas-free prior to scrapping has resulted in numerous deaths and injuries in the Bangladeshi yards.[107] It was only in February 2011 that shipbreaking was recognised as an industry and hence subject to the *Labour Law Act* 2006.

Amongst the defences made by the plot owners in favour of their industry—apart from the high level of employment that it generates—is the environmental argument that it saves the forests from being chopped down by the importation of furniture from the old ships—although a large area of coastal mangrove forest was

---

[100] Zakaria (2011).

[101] The plot owners, however, claim that they supply only 25–30 % of the country's iron demand. Marine News (2010).

[102] There are references to large volumes of steel and non-ferrous metals being smuggled out of the country through the Feni border. Khan (2006), p. 1.

[103] The price of steel billets can rise rapidly—in 2006, the price of steel billets from India was reported at US$430 per ton; after price rises including a 15 % export tax imposed by India, the 2008 price had reached US$1,200 per ton. Basel Action Network (2008).

[104] Greenpeace (2000), p. 4.

[105] International Federation of Human Rights and Young Power in Social Action (2008), p. 7.

[106] Marine News (2010).

[107] On 26.12.2009 cutting work at the Rahim Steel and Shipbreaking yard on the main gas tank of the tanker *Agate* initiated an explosion and a fire that burned for several hours. Seven workers were burned to death, and an eighth was missing and others seriously injured. Kernaghan (2010).

chopped down to provide the sites for the shipbreaking sites—and that they save the banks from huge losses (some 42 ships at the time of the statement that were lying offshore awaiting breaking represented huge government and bank loans).

Off Bangladesh, a vessel for disposal is anchored in international waters in the Bay of Bengal until the various inspections are completed and the vessel is declared gas-free. The sequence of events is:

- The Notification of Cleaning from the State Pollution Control Board, after which the breaker opens a letter of credit
- The vessel arrives at the outer anchorage, from where it proceeds to a berth allocated by the Chittagong Port Authority
- Inspections are to be undertaken by Customs, the Naval Authority and the Technical Committee, the latter submitting its report to the Department of the Environment (DoE)
- The DoE issues clearance to Customs and the CPA, who in turn issue beaching permission to the breaker
- The vessel may then be beached and a request for permission to begin breaking is made to the DoE
- A further inspection is undertaken by the DoE, after which permission may be granted to commence demolition.[108]

The extent and the validity of various inspections have been questionable, in particular, the validity of gas-free certification: hence the large number of fires and explosions at breakers' yards, when workers cut into compartments that contain explosive fumes. Freeing a ship of fumes is an expensive operation for shipowners or shipbreakers. The inspection procedure is reported to be only nominal, with certification easily obtained for money.[109] In particular, the need for gas-free certificates does not appear to be closely regulated, with the result that fires and explosions are regular occurrences. In the 20 years to 2006, some 400 deaths and 6,000 serious injuries have been associated with the yards.[110]

The industry is nominally supervised by both the Ministry of Ports and Shipping and the Ministry of Industries, although numerous other government and non-government bodies may be involved, all operating without any specific co-ordination. The Explosives Department is (nominally) responsible for the issue of gas-free certificates, once the ship has been inspected. All sensitive communications, radar and navigation equipment should be inventoried and confiscated by the Navy. The Mercantile Marine Department list the marine stores on

[108] Jani (2011).

[109] In 2006, the *Alpha Ship*, ladened with asbestos, was given clearance for scrapping at Sitakund, but when the matter was reported in the media, the new owner was forced by the government to send the ship back to the seller. The Daily Star (2006).

[110] Hossain and Islam (2006), p. 1. In December 2009, an explosion in the (uncleaned) tanks of the tanker *Agate* which was under demolition resulted in the deaths of seven cutters, with one further man missing, presumed dead and serious injuries to a dozen more. An amount of Taka 10,000 (US $145) was paid to the families of the dead workers. Bockmann (2010), p. 1.

board, carries out checks of safety measures taken and verifies the vessel's documents. Customs inspect the ship and its cargo and collect the import tax. These steps take about 1 week to complete. Port officials verify import documents and certificates issued by the various government authorities, and verify the payment of duties and taxes. A permit for the ship to enter territorial waters for beaching is finally issued by the Chittagong Port Authority.[111]

Bangladeshi environmental law of 1997 requires that shipbreakers (and other industries) must have an environmental clearance certificate from the Department of the Environment, Ministry of Forest and Environment, which in turn requires the development of an environmental management plan for each shipbreaking site. This requirement went largely unenforced until 2009, when an order from the Bangladesh High Court to re-impose the requirement led to a lock-out by the sites' owners—see Chap. 3. New Bangladeshi rules now require that ships destined for disposal in Bangladesh are accompanied by a certificate from the exporter declaring that the ship is free of toxic substances.[112]

In 2010, the Dutch engineering company GreenDock announced plans for a US $51 million project for a 'green dock wharf' where ships could be dismantled under controlled conditions and hazardous wastes effectively treated—see Chap. 7.[113]

1.4.3 Pakistan

Although Pakistan was the original site of shipbreaking on the sub-continent, the fortunes of the shipbreaking industry in Pakistan have fluctuated with the competition from India and Bangladesh. Although labour costs were the lowest in Asia, rising scrap prices and import duties have led to a progressive decline.[114]

The industry is centred on Gaddani Beach, some 50 km from Karachi and, despite the very small tidal range of just 1–3 m, all but the largest draft vessels can be beached there, the offshore deep water and sandy beach facilitating the beaching. The industry rose to prominence as the world's largest shipbreaking site during the 1980s, when the government announced a series of supportive measures. At that time, some 30,000 workers were directly employed in shipbreaking, all working at the same basic level as elsewhere, with a further 500,000 estimated to be engaged in related activities.[115] Today the workforce is estimated at between

[111] Hossain and Islam (2006), p. 8.

[112] Lloyd's List (2010).

[113] Integrated Regional Networks (2010). IRIN is a non-profit organ of the UN office for the Co-ordination of Humanitarian Affairs (OCHA), and is based in Nairobi.

[114] Putherchenril (2010), p. 30.

[115] Putherchenril (2010), p. 30.

12,000 and 15,000.[116] By the late 1980s, a high import duty of 45 % resulted in a decline of the industry to the extent that only six ships were demolished in 2005, the decline having a particularly hard impact upon the smaller re-rolling mills that were dependent upon the supply of raw materials from the shipbreaking yards.[117] Lobbying of the government brought a reduction of taxes to enable the industry to continue, these measures lowering the landed cost of ship-scrap and enabling the shipbreakers to bid for higher priced ships in the world market.[118] Another revival in the early 1990s was associated with a general fall in the price of ships for scrap on the international market, but those prices rose once more as the supply of ships fell and Pakistani shipbreaking entered another period of decline.[119] Currently there is no import duty on ships for recycling in order to facilitate the growing demand for steel, yet a combination of sales tax, withholding tax and other taxes, results in an effective import tax that is more than double that of India and Bangladesh.[120] As in the other shipbreaking states in the sub-continent, there is a high rate of re-use of fittings taken from the ships.

Since 2008, the yards have again been very active, and in 2011 Pakistan had risen to the third largest site. Today, some 68 plots are active under some 38 operators[121] although two plots are required to beach a ship.[122] The NGO Shipbreaking Platform, however, categorises the legal standing of the industry as '*unclear*,' with few specific regulations in place, such laws as existing being general to the country rather than to the industry and then only weakly monitored and enforced. Asbestos wasted is reported to be merely bagged and then dumped around the yards, there being no specific facilities available for controlled disposal, since unlike as in India, asbestos may not be resold in Pakistan.[123] Since 2011, responsibility for environmental and labour matters relating to shipbreaking have been shared between federal and provincial governments.[124] Inspections of yards and ship are under the remit of the Balochistan Environmental Protection Agency, who are to issue a No Objection Certificate before newly arrived ships may be cleared, but such documentation is reportedly often issued without ships actually being bordered.[125]

[116] Iqbal and Heidegger (2013), p. viii.

[117] Although this fall in supply affected all re-rolling mills in the country, the larger ones were able to continue by using the more expensive iron billets from Pakistan Steel, which were beyond the means of the smaller mills. Lasbela District Govt. n.d.

[118] The Central Board of Revenue (CBR) fixed the price for sales tax calculation at US$300 per LDT instead of invoice or dutiable value, and sales tax on the local supply of the resultant scrap was reduced from 14 to 5 %. These were issued under notifications SRO 77(1) 27.1.05 and SRO 76 (1) 27.1.05 respectively. Pakistan Observer (2005).

[119] Lasbela District Govt. n.d.

[120] Khan (2010).

[121] Iqbal and Heidegger (2013), p. viii.

[122] Iqbal and Heidegger (2013), p. 19.

[123] Iqbal and Heidegger (2013), p. 22.

[124] Iqbal and Heidegger (2013), pp. 16–17.

[125] Iqbal and Heidegger (2013), p. 21.

The sandy shoreline allows cranes and other mechanical equipment to be employed instead of manual handling and, instead of the use of gas cylinders, underground supply lines deliver oxygen and LPG from tanks. Any PPE employed is reportedly recovered from the ships under demolition and its use is very limited—only gloves are reportedly provided for asbestos removal, which takes place in the open.[126] Two trades union are present, the Ship Breaking Labour Union Gadani and the Ship Breaking Democratic Workers Union, although membership is not extensive and the Pakistan Shipbreakers Association does not recognize the latter.

Workers are mainly immigrants from Northern Indi, operating on a no-contract basis; child labour is not employed. Assessments of working and living conditions vary, with Ul-Hasan reporting that conditions are somewhat better than in other sites, with basic free facilities of dormitories and canteens available in the yards[127] although potable water has to be brought in by trucks from the next town. The report from the NGO Shipbreaking Platform, however, presents a picture of workers living in shanty towns made from plywood taken from the ships, and an absence of schools, medical facilities proper sanitation, a public supply of clean drinking water and few food shopping facilities. Work is 7 days a week and health care and emergency provisions on site are minimal—the nearest hospital is in Karachi.[128]

The effective closure of the Bangladeshi yards by order of the High Court during 2010–2011 gave a boost to the Pakistani breaking industry, with over 100 ships being bought by the Gadani Beach yards in 2010. Prominent amongst these were the tankers and gas carriers—the wet tonnage—that would traditionally have gone to Bangladesh.[129] Like Bangladesh, Pakistani breakers tend to prefer the larger ships for breaking, as they generally represent a relatively simple construction.

1.4.4 China

In common with other Asian shipbreaking states, China has experienced fluctuations in its fortunes, but is currently experiencing a major growth in its industry, due mainly to the differing conditions under which it operates. In the 1980s, the increase in taxation on imported tonnage for scrap almost eliminated China from the world market,[130] but by 1993 nearly half of the ocean-going ships sent for demolition again went to China, making it the then world leader in shipbreaking, after which time the introduction of new taxes caused China to slip once more in the ranking.[131]

---

[126] Iqbal and Heidegger (2013), p. xi.

[127] Ul-Hasan (2011).

[128] Iqbal and Heidegger (2013), p. x.

[129] McCarthy (2010b), p. 4.

[130] Andersen (2001), p. 3.

[131] Greenpeace (2001), p. 5.

The industry's fortunes rose in the later 1990s and early 2000s, before sliding back in the period 2005–2008, owing to the boom in the shipping industry and the consequent scarcity of ships available for scrapping. China's fortunes rose again with the severe downturn of the shipping industry's fortunes in 2009, together with a series of measures by the State Council to encourage local shipping companies to replace ageing vessels to aid the national shipbuilding yard, plus a strict control by the Chinese government of the export of Chinese ships for scrap.

China currently has some 90 yards situated in the lower reaches of the Yangtze and Pearl Rivers. The main centre is located at Zhang Jiang in Jiangsu province, some 220 km from the sea on a 1,500 m canal off the river; four ships can be broken simultaneously. Other sites are located in Guangdong and Fujian provinces. The yards tend to be situated on small plots of land, with agriculture practised immediately adjacent to the sites and no transitional industrial areas between the areas of dense population, just a few kilometres away. Ships are broken up alongside quays, rather than on the beach; this makes it easier for ships to be delivered to the breaking site for demolition.[132] In this way, ships are demolished in a manner that mirrors the reverse of the shipbuilding process, rather than being cut up in chunks or in slices as in the beaching method. Cranes on the quays are available for lifting the scrap sections. Like Bangladesh, Chinese shipbreakers have a preference for the large VLCCs and ULCCs. Ships are broken up whilst still in the water and, although oil booms were employed, they proved to be inadequate to prevent the spread of oil pollution at that time.[133] In the first of Greenpeace's visits to the Chinese yards in 1999,[134] the handling of asbestos was seen as a major problem; with regard to one ship being demolished, the former owners (Hamburg Süd) had provided a detailed inventory of the types and locations of asbestos materials on board a ship observed, and had provided protective suits for those employed in asbestos removal, however the inventory appeared to have gone unobserved, with asbestos waste strewn around the site and workers using protective gear working alongside unprotected workers.[135] At the second visit, asbestos removal was still a major problem. As in India, asbestos removed was allegedly separated for resale, although this reuse was officially banned.[136]

Although the control of environmental pollution may still be improved, the conditions under which the actual workers operate are deemed to be somewhat better than those experienced in the sites of the sub-continent. This is due in no small part to the system under which a number of ships have been broken up under direct contract with the shipowners Hamburg Süd, and subsequently with P&O Nedlloyd (later part of the Maersk group). Under this scheme, the shipowners began providing training and protective equipment for the workers and encouraging

[132] Jones (2007).

[133] Jones (2007).

[134] Greenpeace (1999b).

[135] Greenpeace (1999b), pp. 9–10.

[136] Greenpeace (2001), p. 11.

measures of environmental control. This, in its turn, was the result of pressure put upon the companies by Greenpeace as part of its campaign against the traditional conditions of the industry as a whole. This shipowner/shipbreaker partnership is seen as the basis for a more progressive and mutually advantageous form of development, and a number of European organisations are now offering a service to shipowners whereby their vessels can be demolished in specific Chinese yards under the supervision of inspectors with the powers to halt the process, should this be deemed not to be in accordance with predetermined standards—see Chap. 7. Whilst environmental reform may still have some considerable way to go, the industry may be viewed as rather more 'high tech' than the yards of the sub-continent, and thereby, more 'respectable.'[137] However, the then head of Maersk's ship recycling programme was quoted as saying that there are only three or four facilities '*that can do a green job*.'[138] The more recent tightening of environmental and safety laws by the Chinese authorities has not halted the industry's progress, although it may have detracted somewhat from its profitability. This is an industry that has little or no informal sector, but is regarded as a legitimate industry by the government of China.[139] In discussions with one who facilitates shipbreaking in China on behalf of shipowners, the rigid structure of the Chinese government was described as being based upon a pyramid structure of Beijing/province/city model in which decisions from the top are quickly implemented downwards,[140] the role of the Environment Ministry having risen significantly nearer the top of the structure than was previously the case.

The growth of the Chinese industry appears to be largely at the expense of Alang. China's enormous appetite for steel has at times forced up the price of scrap ships to around US$400 per LDT, rendering the operations at Alang less financially viable. Whilst India, which relies more on the import of steel scrap rather than the use of steel from demolished ships, was raising the customs duty on these ships from 5 to 15 %, and reducing the import duty on steel from 25 to 15 %, the Chinese experience was just the reverse, with an import duty on the ships at just 3 %, together with a 17 % value-added tax refund to its breakers. Better demolition techniques in China also result in a better quality of scrap than in the Indian yards,[141] but the costs associated with the safety and environmental management employed means that the 'greener' yards can only offer prices to owners that are US $50–60 less per LDT.[142]

At the beginning of 2010, the Ministry of Commerce, the National Development and Reform Commission, the Ministry of Industry and Information Technology and

[137] Jones (2007).

[138] McCarthy (2010a), p. 4.

[139] McCarthy (2010a), p. 4.

[140] Nevertheless, facilitating agents acting on behalf of other international shipowners make their arrangements directly with the shipbreaking yards.

[141] Basu n.d.

[142] McCarthy (2010a), p. 4.

other related departments jointly announced a draft set of legal guidelines for the industry, aimed at strengthening environmental standards at the yards by tightening monitoring in order to control pollution. Local maritime, fisheries and environmental authorities were to be given regular access to the yards for the purposes of inspections. Older yards were to be phased out and remaining yards prohibited from scrapping ships on the beach. Examination by China's Customs and Environmental Department is part of the greater controls over foreign ships imported for scrapping. Price bargaining and environmental assessment for imports would be formalised by agency services acting between shipbreaking yards and foreign owners. In addition, a system for shipping lines to phase out old ships was to be introduced and older ships were to be banned from operating; classification societies would no longer be able to certify them.[143]

In July 2010, the above organisations announced a subsidy of some Yuan1,500 (US$220) per tonne for Chinese shipowners who scrap older ships and replace them with new builds.[144] A reaction from a representative of the Chinese Shipowners Association suggested that whilst the policy might seem attractive to owners of small fleets, it was likely to have little effect on the larger shipping lines, which already had established replacement policies. Although the policy should have an impact on the ship recycling industry, proper regulations are an *a priori* requirement for this to be effective.[145]

Although most yards in China claim to be certificated to the ISO 30000 or ISO 14001 systems, some consider this to be merely a marketing tool, one inspector allegedly covering some 56 shipbreaking yards in 30 days.[146] Although ISO 14001 is approved by the Chinese Standards Bureau in China, the recycling facilitator interviewed is pushing for external Lloyd's Register/Germanischer Lloyd-style audits based on the facilitator's own practices. Turkey allows LR and GL, so ISO 14001 is recognised as having greater value there.[147]

A major development of state-of-the-art ship building, repairing and recycling facilities is currently under way at Dalian, and will add considerably to the

[143] Ching-Hoo (2010a), p. 10.

[144] The scheme was to operate until the end of June 2012, and was accompanied by a range of conditions, including the fact that ships eligible must be registered in China, of defined tonnage and trading history. Ships scrapped must be replaced by an equal or greater amount of tonnage to be built in Chinese yards and licensed to operate both domestically and internationally. Ching-Hoo (2010b), p. 5. Similar schemes have in the past been operated by other states such as the UK and Japan.

[145] Ching-Hoo (2010b), p. 5.

[146] Compare this with the author's own experience in obtaining ISO certification for a multinational company with dedicated environmental management staff; the ISO 14001 survey for a single head office site (largely office activities) took some 5 man days, whilst the ISO 9001 inspection of a large automotive parts warehouse required some 11 man-days.

[147] Comment from recycling facilitator with close knowledge of Chinese yards, made during discussions 2.5.2012. It is also the view of the author, with professional experience of international standards, that the majority of similar certifications have little inherent value to the extent that ISO envisaged.

shipbreaking capacity of China—see Chap. 7.[148] To give some perspective—the shipbreaking capacity of 1 million LDT at China's operations at Jiang Xiagang and at the new facilities under construction by Dalian is each considered to be some seven times that of the capacity currently existing in the OECD.[149]

1.4.5 Turkey

Shipbreaking in Turkey has a somewhat singular position in the shipbreaking world in that Turkey is the one nation amongst the major breakers that is actually a member of the OECD, and hence in the strict letter of the laws relating to the export of hazardous waste, is the one destination to which European countries subject to the European *WSR* and other signatories to the *Basel Convention*[150] should send their ships for scrapping if they are not to be demolished at home or within Europe. Turkey is also the largest importer of steel scrap in the world and amongst the top ten producers of steel products.

During the 1950s and 1960s, a few companies were breaking ships near Istanbul, with little regulatory control, but the negative environmental impacts led the government to call a halt to the activities at the beginning of the 1970s. By 1974, the demolition of ships had restarted at Aliağa, Izmir, which was established as a ship recycling area by government decree in 1976, and legalised as an industry in 1986 by the issue of government regulations on ship dismantling.[151] At that time, Turkey had risen to third place in the dismantling rankings, after South Korea and Taiwan. Today, Turkey holds fourth place after India, Bangladesh and China. Its relatively modest size when working at full capacity provides work directly for some 2,800 workers, with a total of 8,000 including the associated industries. Aliağa's geographical location is attractive to owners of vessels finally discharging in Europe, since the costs of voyages in ballast to the sub-continent, plus Suez Canal fees, are obviated.[152]

In 1995, Turkey incorporated the principles of the *Basel Convention* into its *Regulation to Control Hazardous Wastes*.[153] Nevertheless, a Greenpeace report of 2002[154] on their inspection visit to Aliağa painted a picture of practices at that time

[148] Feng (2011).

[149] Blanco (2012). Soledad Blanco is the Director of the Sustainable Resources Management Directorate, Industry and Air Department, Environment Directorate General, European Commission.

[150] *Regulation 1013/2006/EC on the shipment of waste*, applicable since 12.7.2007.

[151] Ship Recyclers' Association of Turkey (2007).

[152] Lloyd's List (2012a), p. 8.

[153] 27.8.1995, No. 22387.

[154] Greenpeace (2002b).

were not significantly different from the yards of Asia, with asbestos spread across the sites, wastes being burned on open fires etc. A number of other regulatory measures designed to prevent environmental pollution were available to the Turkish authorities, but were simply not adequately enforced.[155] Over recent years, much effort has been applied to making the operations environmentally more effective and an inter-ministerial scheme for licensing the industry was introduced in 2006. Dangerous work permits are issued by municipalities to those facilities which meet the required standards. A licence for the removal and temporary storage of asbestos and PCBs was issued to the Ship Recycling Association of Turkey by the Ministry of Environment and Forestry in 2007.

Turkey now has a record of banning particularly hazardous ships from entering its waters for demolition—see, for example, Sect. 5.2, Chap. 5. State bodies control and audit the rules governing the industry, and training programmes in occupational health and safety and hazardous materials handling are promoted by the industry's own national ship recycling association. Health checks are periodically made and records held for inspection by labour inspectors.[156]

Turkey employs a combination 'afloat and landing' method of demolition by which ships are progressively pulled ashore on to concrete ramps as demolition progresses, this method providing a measure of containment and collection for the spillages that pollute the environments of other shipbreaking nations. If the after part of a vessel (that is not yet being cut) remains in the water, oil booms are deployed to prevent pollution. Smaller ships (below 1,000 LDT) are totally demolished ashore.

Waste control in Turkey is governed by two overarching pieces of legislation, namely the *Regulation on General Principles of Waste Management* (2008) and the *Regulation on Solid Waste Control* (*1991 amended 2005*). A single company is employed by all the yards collectively to remove asbestos waste safely, and government has invested in facilities for the safe removal of other wastes from vessels during dismantling, the transport of waste being undertaken by a company licensed by the Ministry of Environment.[157] A pilot project to recycle up to 29,000 tons hazardous waste by gasification was opened at Kemerburgaz, Istanbul in 2009. Sea water samples are examined periodically for traces of pollution and cranes inspected at 3 monthly intervals.[158]

The long term leases on the 25 plots facilitate investment in provisions and in infrastructure; however, the higher wages and overheads of the Turkish yards, compared to sites on the sub-continent, mean that prices paid to owners are

[155] For example the *Regulation for Aliağa Shipbreaking Yards*, in force initially 1970 and revised for the second time in 1997 after an explosion on board a tanker being demolished killed some seven workers; *Regulation on Hazardous Chemicals* 10.7.1993; *Environmental Law no. 2872*, 9.8.1983; *Regulation on Workers Health and Occupational Safety* 11.1.1974.

[156] Ship Recyclers' Association of Turkey (2007).

[157] McCarthy (2010c), p. 4.

[158] Cevike (2010).

below those offered elsewhere, and the industry is therefore somewhat under-utilized. Despite the higher costs resulting from the more regulated regime, Turkey remains the one major shipbreaking nation that meets one of the conditions for handling hazardous waste under the provisions of *Basel*. Entry into the EU that Turkey is seeking would bring the industry further into the realms of the EU strategy for shipbreaking. In 2009, 127 ships, representing 300,000 LDT, were scrapped in Turkey, against an annual maximum capacity of 900,000 LDT.[159] In July 2010, a delegation from Bangladesh visited yards in Turkey to learn how their own industry might be improved.[160]

1.4.6 Other

1.4.6.1 USA

In the USA, shipbreaking is now a relatively minor industry, attending mainly to the redundant ships from the US Navy and the US Maritime Administration (MARAD), the latter being responsible for former naval auxiliary vessels, many of which made up the ageing US Reserve Fleet,[161] which is finally being scrapped. For a period in the 1970s, the industry was forced to close, and until recently just six sites represented the remnants of the former industry of some 100 sites.[162] In 2011, the first recycling site on the West coast was opened at the former Mare Island Naval Shipyard, California, thereby obviating the need for decrepit vessels from the reserve fleet at Suisun Bay, San Francisco to make the 5,000 mile trip via the Panama Canal to East coast or Gulf coast breakers.[163] The dire conditions observed at Brownville, Texas, prompted the publication of articles in the *Boston Sun* in 1997 that initiated the now international concern over the state of the industry worldwide. However, environmental protection controls are now high—the US Environmental Protection Agency's (EPA) *Guide for Ship Scrappers*[164] being issued in 2000 and dismantling is overseen by MARAD and Navy Sea representatives. It is now US policy not to export warships for demolition; however, merchant ships invariably still go to the yards of Asia. The Brownsville ship canal just three miles from the Mexican border remains the major centre of the US shipbreaking industry.

---

[159] Gorkem and Kemerci (2010).

[160] McCarthy (2010c), p. 4.

[161] The US Reserve Fleet was begun after WWII and consisted of ships kept in preservation as a strategic reserve and distributed in sites at Suisan Bay, California, Beaumont, Texas, and in the James River, Chesapeake Bay.

[162] Nijkerk (2005).

[163] The Maritime Executive (2011).

[164] Environmental Protection Agency (USA) (2000). The guide was issued to provide site supervisors '*with a good understanding of the most pertinent federal environmental and worker safety and health requirements affecting ship scrapping/breaking operations*.'

In 1950, the combined US Reserve Fleet numbered over 2,000 ships, subsequent legislated disposals being prompted by the deteriorating conditions of the ships and the potential environmental pollution they represented.[165] In 2010 it was determined that, prior to being moved for scrapping, the ships are to enter local dry docks for the removal of invasive species.[166] Reefing has been a popular option for a number of US naval vessels, and ships as large as the former aircraft carrier *Oriskany* have been deliberately sunk after extensive stripping of hazardous materials involving many months of work. To ensure that environmental standards are maintained, the government usually has to pay considerable sums to the domestic breakers; as an example, the cost of cleaning and demolishing two cargo ships from the reserve fleet in 2009 cost the US government a total of US$3.6 million, or US $409 per ton.[167] In addition, there are instances in the past where the authorities have retaken ownership of vessels when the breaking has not been to the required standards.

In 2010, proposals were issued by a company in the Grand Marianas for the demolition of MARAD ships, particularly those in the Suisan Bay fleet, in Saipan, which being a US commonwealth state would not violate US laws restricting the disposal of such ships to US yards.[168]

1.4.6.2 UK

Shipbreaking in the UK was formerly a large industry, with many small sites around the coast. After 1945, large numbers of naval vessels were rendered surplus to requirements and quickly found their way to the breakers' yards. The subsequent reduction of the fleet, plus the decline in the merchant navy, and the development of health and safety and environmental restrictions saw the UK shipbreaking industry reduce almost to extinction. Some revival was aided by the need to break up obsolete platforms from the new offshore energy industry by companies such as Able UK, based at Hartlepool on Teesside. In 2003, Able UK secured a contract to import some 13 surplus and decaying naval auxiliaries (the 'Ghost Ships') from the US Reserve Fleet for disposal at Hartlepool. An intensive campaign by activists[169]

---

[165] Centre for Land Use Interpretation (2010).

[166] Centre for Land Use Interpretation (2010).

[167] The *Pan American Victory* and *Earlham Victory* were taken from the Suisan Bay Reserve Fleet, cleaned of organisms to prevent the transfer of invasive species and towed to Brownsville, Texas for demolition. Robin des Bois (2010b).

[168] Contra Costa Times (2010).

[169] Perhaps a little ironically, Able UK were actually supported in their intentions by Greenpeace, opposition coming from Friends of the Earth, who campaigned against the anticipated further pollution of an already blighted area in the absence of any assessment of the potential environmental and economic threats that the scrapping of these ships might represent. The site for the disposal was adjacent to Seal Sands, a Site of Special Scientific Interest (SSSI) protected under European and international law.

resulted in the number of ships actually arriving at Teesside being reduced to just four after a successful action[170] in the DC Federal District Court in 2003 and the granting of an injunction by the High Court in London requested by Friends of the Earth (FoE). The import of the American ships was followed by the award of the contract to Able UK in 2008 to scrap the former French aircraft carrier *Clémenceau* (see Chap. 5). It also initiated a period of licensing of sites for (relatively modest) shipbreaking operations in the UK at sites such as the former shipbuilding yard of Harland and Wolff.

The 'Ghost Ship' controversy became the subject of a hearing by the House of Commons Select Committee on Environment, Food and Rural Affairs, in November 2003, and the following year the Committee issued a report calling for the development of an industry that could dismantle ships to high standards at home rather than rely on their export to Asia, setting an example with the disposal of Government-owned ships in the UK. Any government vessels sold for scrapping would be subject to the conditions of recycling according to *Basel* principles, only in OECD countries and with the new owners—*i.e.* the breakers—taking full responsibility for safe and environmentally sound recycling.[171] Following the *Ballard Report*[172] and the Select Committee report,[173] DEFRA issued a draft guidance to the UK ship recycling strategy in March 2006, followed by the *UK Ship Recycling Strategy, Guidance, and a Regulatory* impact assessment in February 2007.

At the end of 2008, the former assault landing ship *H.M.S. Intrepid* arrived at Liverpool for scrapping, the largest Royal Navy ship to be broken up at home for a number of years. The *Intrepid* was followed by the naval auxiliary tanker *Grey Rover* and the landing ship *Sir Percivale*. However this UK revival may have been no more than a temporary or even token matter, with the Select Committee's recommendation that government-owned ships be broken up at home having a somewhat limited shelf life, since in recent years, Turkish shipbreakers at Izmir have demolished the former aircraft carriers *H.M.S. Invincible* and H.M.S. *Ark Royal*, six Type 42 destroyers and three auxiliary tankers from the

[170] The action, filed in September 2003 by Earthjustice on behalf of the Basel Action Network and the Sierra Club, was based upon the claimed violation of the US Toxic Substances Control Act (TSCA), which prohibits the export of substances containing PCB. According to Earthjustice, the US government's own estimates of hazardous materials on board these 13 ships included 700 tons of PCB, 1,400 tons of asbestos and over 3,000 tons of waste fuel oil. The export of these vessels had been seen as a test-run for the export of larger numbers of these obsolete naval and auxiliary ships to developing countries for disposal. Basel Action Network (2004).

[171] With regard to ships sold for further reuse, the new strategy document required that the government would negotiate the inclusion of similar provisions, not stipulated, as had been included in the draft strategy. ENDS (2007), p. 44.

[172] Ballard (2004).

[173] House of Commons Environment, Food and Rural Affairs Committee (2004).

Royal Navy, with a further three Type 22 frigates scheduled for departure in the near future.[174]

More recently, ship recycling facilities have been licensed within the dry docks at Swansea. Application to the Environment Agency for a permit to operate was made under the provisions of the *Environmental Permitting (England and Wales) Regulation 2010* and began in February 2010 and took some 2 years to complete, the first ship arriving for demolition just 2 years later. As well as the dismantling of the ship, activities include the handling and storage of wastes, including hazardous waste, these being classified as recovery rather than disposal operations and guidance was under the EA's general guidance for the '*Recovery and Disposal of Hazardous and Non Hazardous Waste*,' there being no specific sector guidance to follow.

The process of permitting at Swansea was particularly challenging for the operators; the simple transition of classification for a ship from the subject of a repair operation to a waste operation resulted in many conditions and restrictions being imposed that were not required for the ship repair operations carried out at adjacent berths.[175] The hulls of ships painted with TBT can be cut but the TBT itself may not be removed from the structure. Strenuous steps are taken to prevent contamination by invasive species. Ballast water may not be discharged within the dock and all such water, including any water falling anywhere on the facility has to be collected in large attenuator tanks and regularly sampled prior to any discharge, Special bunds had to be constructed between each ship and the dock gates to control leakage. Numerous technical assessments were demanded on the potential impacts of noise; odors; drainage; accidents; environmental risks, etc. to answer many public objections and a detailed public consultation and due diligence exercise called for. Ultimately, the first ship for disposal by Swansea Drydocks Ltd. arrived on 1 February 2013 and was recycled by the end of the following month; no complaints were received. The ex-Royal Navy frigate *HMS Cornwall* is scheduled to arrive for demolition in the final quarter of 2013.[176]

Finally, one case of demolition after unintended beaching that took place in the UK. On the evening of 31 January 2008, a severe storm broadsided the ro–ro ferry *Riverdance*, causing her cargo to shift and the vessel to list, the ferry finally being driven onto the beach at Clevely, near Blackpool and finishing up on her side on dry sand at low tide. All efforts to refloat her over 2 months failed and it was decided to break the vessel up in situ, work being undertaken in consultation with the Environment Agency. The engines, cargo and 150 tonnes of fuel oil were removed, leaving just engine oils as remaining potential pollutants; an oil spill contractor was

---

[174] Ships Monthly (2013b), p. 15.

[175] As an example, ships are not allowed to be stored in the dock's wet berths, but must immediately be taken into dry dock on arrival.

[176] Cumberlidge (2013).

on standby. Since no environmental permit for the operation existed, the EA advised that Regulation 40 of the *Environmental Permitting Regulations*[177] applied, to prevent damage during what was an emergency situation.[178] The scrap generated during the demolition was temporarily stored on the promenade in a secure compound. The wreck was broken up in a few months, no oil pollution occurred, and although some containers initially spilled their loads no debris remained on the beach. Whilst it is readily acknowledged that the *Riverdance* was certainly not of the proportions of many of the ships currently demolished, the grounding (as distinct from deliberate beaching) was entirely unexpected and unprepared for, yet there was no impact upon the environment.

1.4.6.3 Others

In addition to the major shipbreaking sites detailed above, a large number of smaller sites are in operation in many countries around the world. Europe's smaller yards around Denmark, Norway, Poland, Netherlands and Lithuania have benefitted from the shipping turndown as local short sea and fishing operators turned to local yards to offload their redundant tonnage, thereby improving on disposal savings that even Turkey could offer.[179]

Whilst not attracting the attention given to other breaking sites, shipbreaking in Denmark is an important industry for that country. The Fornæs yard in the port of Grenå recycled 22 ships in 2012 and although the total tonnage represented may be small compared to other sites around the world, it makes this one of the largest sites in Europe. Ships are stripped down and engines removed whilst the vessels are still in the water, all hazardous materials being handled by a duly certified third party and the hull is then pulled onto a slipway (representing some €3 million investment), which incorporates new containment provisions to capture any subsequent discharges. Operations at the yard are overseen by the Danish Environment Agency.[180] In the former Maersk yard in Munkebo, three ex-Danish Navy corvettes arrived to be broken up in dry dock early in 2013.

A small but thriving shipbreaking industry operates in Indonesia, working under conditions similar to those at Alang but overshadowed by the size of the latter. In 1994, the Japanese shipbuilding company Tsuneishi Heavy Industries set up an industrial zone of West Cebu Industrial Park (WCIP), in the West Cebu Island in the Philippines, to break up ships. WCIP represented an area where both labour

[177] 2007 *Environmental Permitting Regulations SI 2007/3538.*

[178] The wreck attracted thousands of visitors and it was assumed that in the absence of a permit, the Secretary of State's Representative (SOSREP) would have followed the same decision for safety reasons—comments received in correspondence with a member of the Preston office of the EA who was involved.

[179] Lloyd's List (2012b), p. 6.

[180] NGO Shipbreaking Platform (2013).

costs and regulation levels were low. Ships were reportedly dismantled in the sea with little effort to restrict pollution.[181]

Attempts to initiate a shipbreaking industry on the island of Bolama in Guinea Bissau were made by a Spanish company in 2002 under the guise of major job creation in fishing, agriculture and tourism sectors. These plans were countered by an information campaign mounted by the Forest and European Union Resources Network (FERN), which pointed out the inherent dangers to the ecosystem upon which the population depended and resulted in the successful abandonment of the scheme.[182]

References

Abdi R (2003) India's ship-scrapping industry: monument to the abuse of human labour and the environment. IIAS Newsletter, 32 (November 2003)

Agrawal S (2011) Cash-buyer woes. Tradewinds, 20 April 2011

Andersen AB (2001) Worker safety in the ship-breaking industries: an issues paper. International Labour Organization

Ansari Z (2012) Commercial Director, GMS. In: 7th annual ship recycling conference, June 2012, London

Ballard J (2004) US naval ships: review of regulatory structure. Environment Agency, Bristol

Bari D (2006) In the shipyards of Alang, they destroy men as well as ships. L'Humanité, 28 February 2006, Paris

Basel Action Network (2004) US District Court of Columbia, *Basel Action Network, a Sub-Project of the Tides Centre and Sierra Club v. MARAD and USEPA*. Plaintiffs' motion for summary judgement Case No. 03-02000, 6 August 2004. www.ban.org/library/04_06_01_MOTION_AND_MEM0_FOR_SJ.pdf. Accessed 2 Jan 2011

Basel Action Network (2008) Scrap ships crowd Bangladesh shore on demand for cheap steel. BAN Toxic Trade News, 17 August 2008. www.ban.org/ban_news/2008/080817_crowd_shore_on_demand_for_cheap_steel. Accessed 1 June 2010

Basu I (n.d.) Indian ship-breaking drifts to China. Asia Times Online (n.d.). www.atimes.com/atimes/South_Asia/FD22Df05.html. Accessed 18 Jan 2010

Bhattacharjee S (2009) From Basel to Hong Kong: international environmental regulation of ship-recycling takes one step forward and two steps back. Trade Law Dev 1(2):193

Blanco S (2012) EU regulatory review. In: 7th annual ship recycling conference, 14–15 June 2012, London

Bockmann MW (2010) Report condemns Bangladesh yards. Lloyd's List, 13 January 2010, London

Central Technical Team (India) (2009) Report of the Central Technical Team constituted by Ministry of Environment & Forests on the inspection of ship 'Platinum-II' anchored at Bhavnagar anchorage point. CTT, 23 October 2009

Centre for Land Use Interpretation (2010) American ship breaking: It all comes apart at the bottom of America. The Lay of the Land Newsletter, Spring 2010. www.clui.org/lotl/v33/k.html. Accessed 11 June 2010

[181] Furuya (1998).

[182] Forest and the European Union Resources Network (2008).

Cevike E (2010) Ship Recycling Association of Turkey. Ship Recycling Technology and Knowledge Transfer Workshop, Izmir, 14–16 July 2010
Ching-Hoo H (2010a) China unveils tougher legal guidelines for shipbreakers. Lloyd's List, 3 March 2010, London
Ching-Hoo H (2010b) Doubts over China scrapping scheme. Lloyds List, July 2010, London
Contra Costa Times (2010) Ships from polluting mothball fleet in Suisan may go to South Pacific. http://www.contracostatimes.com/ci_14735685?nclick_check=1. Accessed 25 Mar 2010
Cumberlidge P (2013) Case study: Swansea Dry Docks; Wales' first ship dismantling facility; setting the blue print for ship recycling in the UK. In: 8th annual ship recycling conference, 26–27 September 2013, London
Det Norske Veritas (2009) DNV interim guidelines on lay-up of ships. www.dnv.com/industry/maritime/publicationsanddownloads/publications/updates/bulk/2009/guidelineforthelayupofships.asp. Accessed 10 Feb 2010
Drury S (2011) Legal minefield or land of opportunity: potential legal pitfalls in sales of end of life vessels and how to avoid them. In: 6th annual ship recycling conference, 14–15 June 2011, London
Dube HC (n.d.) Environmental development in coastal regions and small islands: making unsustainable development sustainable: the case of the Alang ship-breaking Industry in Gujarat, India. www.unesco.org/csi/pub/info/wise4.htm?. Accessed 8 Mar 2010
ENDS (2007) Ship recycling strategy sets minimum standards. ENDS Report 386, March 2007
Environmental Protection Agency (USA) (2000) A guide for ship scrappers. Tips for regulatory compliance. EPA 315-B-00-001, EPA, 2000
European Commission (2004) Oil tanker phase out and the ship scrapping industry. Report. P-59106-07 Final
Feng G (2011) China's Dalian modern ship recycling plant investment. In: 6th annual ship recycling conference, 14–15 June 2011, London
Forest and the European Union Resources Network (2008) Provoking change. A toolkit for African NGOs
Furuya S (1998) Japanese shipbreaking breaks workers in the Philippines. Japan Occupational Safety and Health Resource Centre JOSHRC, 1998. www.jca.apc.org/joshrc/english/14-2.html. Accessed 25 June 2010
Galley M (2013) Hazardous materials in shipbreaking – where do the liabilities lie? PhD research thesis, Southampton Solent University, Southampton
Gorkem A, Kemerci B (2010) Overview of the Turkish ship recycling industry, the main actors and national legislation in Turkey. In: Ship Recycling Technology and Knowledge Transfer Workshop, 14–16 July 2010, Izmir
Greenpeace (1999a) Ships for scrap. Steel and toxic wastes for Asia. A fact-finding mission to the Indian shipbreaking yards in Alang and Bombay. Greenpeace, Hamburg
Greenpeace (1999b) Ships for Scrap II. Steel and toxic waste for Asia. Worker health and safety and environmental problems at the Chang Jiang ship-breaking yard operated by the China National Shipbreaking Corporation in Xiagang near Jiangyin. Greenpeace, Hamburg
Greenpeace (2000) Shipbreaking: a global environmental, health and labour challenge. A Greenpeace report for IMO MEPC 44th Session, March 2000
Greenpeace (2001) Ships for scrap IV. Steel and toxic wastes for Asia. Findings of a Greenpeace visit to four shipbreaking yards in China, June 2001. Greenpeace, Netherlands
Greenpeace (2002a) Towards green ships Ship-breaking International Maritime Industries Forum, October 2002, London
Greenpeace (2002b) Ships for scrap V. Steel and toxic wastes for Asia. Greenpeace report on environmental, health and safety conditions in Aliağa shipbreaking yards, Izmir, Turkey. Greenpeace Mediterranean
Greenpeace (2003) Ships for scrap VI. Steel and toxic wastes for Asia. Findings of a Greenpeace visit to Darukhana shipbreaking yard in Mumbai, India. December 2002, Greenpeace, Netherlands

Gujarat Maritime Board (2007) Salient features of Ship Recycling Regulation 2006. www.gmbports.org/Alang_salientfeature.htm. Accessed 3 Jan 2010
Hossain MM, Islam MM (2006) Ship breaking activities and its impact on the coastal zone of Chittagong, Bangladesh: towards sustainable management. Young Power in Social Action
House of Commons Environment, Food and Rural Affairs Committee (2004) Dismantling defunct ships in the UK. 2003–2004 HC834
Hunter G (2011) Recyclon, a new standard contract for the 'Green' recycling of ships. In: 6th annual ship recycling conference, 14–15 June 2011, London
I-Maritime Consultancy (2001) Ill equipped for shipbreaking? www.imaritime.com/resources/monitor/Archives/Feb-May2001/Feb-May200_shipbreaking.asp. Accessed 11 Nov 2004
Ibeanu O (2010) Report of the Special Rapporteur on the adverse effects of the movement and dumping of toxic and dangerous products and wastes on the enjoyment of human rights. Addendum Mission to India UN General Assembly, Human Rights Council Fifteenth Session Agenda item 3 A/HRC/15/22/Add.3, UN, 2 September 2010
Infochange India (2011) One in six Alang ship workers has asbestosis: Govt Report. www.Infochangeindia.org/environment/news/one-in-six-alang-ship-workers-has-asbestosis. Accessed 11 Nov 2011
Integrated Regional Networks (2010) Bangladesh: taking toxins out of ship-breaking. www.irinnews.org/Report/90376/BANGLADESH-Taking-toxins-out-of-ship-breaking. Accessed 15 Apr 2012
International Chamber of Shipping (2012) Shipping and world trade. www.marisec.org/shippingfacts/worldtrade/number-of-ships.php. Accessed 1 Apr 2012
International Federation of Human Rights (2012) Investigative mission: where do the 'floating dustbins' end up? Labour Rights in Shipbreaking Yards in South Asia. The Case of Chittagong (Bangladesh) and Alang (India). Report no. 348/2, FIDH
International Federation of Human Rights and Young Power in Social Action (2008) Childbreaking yards. Child labour in the ship recycling yards. FIDH and YPSA in association with The International Platform on Shipbreaking, 2008
International Maritime Organization (2003) IMO Guidelines on ship recycling. IMO, London. Adopted at the International Maritime Organization Assembly 23rd session. Resolution A.962 (23) 5.12.2003, Item 9.7 International Metalworkers' Federation 2003. www.imfmetal.org/main/index.cfn?n=47&1=2&c=8238. Accessed 22 Aug 2008
International Metalworkers Federation (2003) www.imfmetal.org/main/index.cfn?n=47&1=2&c=8238. Accessed 22 Aug 2008
International Metalworkers' Federation (2007). Status of shipbreaking workers in India – a survey. Labelled 2004–7, but with a preface dated March 2006
Iqbal KMJ, Heidegger P (2013) Pakistan shipbreaking outlook: the way forward for a green ship recycling industry – environmental, health and safety conditions, 1st edn. Sustainable Development Policy Unit and NGO Shipbreaking Platform, Islamabad/Brussels
Jani D (2011) Ship recycling SWOT analysis for India and Bangladesh. In: 6th annual ship recycling conference, 14–15 June 2011, London
Jones SL (2007) A toxic trade: shipbreaking in China. A China environmental health project fact sheet. Western Kentucky University, Kentucky
Joshi V (1999) Industrial safety concerns in the ship breaking industry/Alang-India. Wise Coastal Practices for Sustainable Human Development Forum, 1999. www.csiwisepractices.org/?read=280. Accessed 21 July 2004
Khan MA (2006) Sea polluted under authorities' nose. The Daily Star, 31 July 2006, Dhaka
Khan AA (2010) Ship recycling in Pakistan. In: Ship Recycling Technology and Knowledge Transfer Workshop, 14–16 July 2010, Izmir
Kernaghan C (2010) Eight more workers burned to death in Bangladesh shipbreaking yard. www.nicnet.org/alerts?id=0014. Accessed 2 Feb 2010
Kimmins S (2012) Sustainable shipping initiative closed loop materials management. In: 7th annual ship recycling conference 29–30 June 2012, London

Lasbela District Govt. (n.d.) Gadani ship-breaking. www.lasbeladistrictgovt.com/gaddani/ship%20breaking.com. Accessed 3 Sept 2008

Lloyd's List (2010) Europe urged to adopt green scrapping. Lloyd's List, 2 March 2010, London

Lloyd's List (2012a) Shipbreakers cash in on prime location. Lloyd's List, 19 July 2012, London

Lloyd's List (2012b) North Europe's small recycling yards ride growth in demand for scrap. Lloyd's List, 10 August 2012, London

Marine News (2010) Ship breaking: points, counterpoints and a focal point. http://marine-news.net/Shipbreaking_Points_counterpoints_and_a_focalpoint-i1517. Accessed 17 June 2010

McCarthy L (2010a) Breaking industry in tug-of-war over environmental standards. Lloyd's List, 2 July 2010, London

McCarthy L (2010b) Bangladesh scrap ban prompts regional shift. Lloyd's List, 15 December 2010, London

McCarthy L (2010c) Turkey proves it is possible clean up a dirty business. Lloyd's List, 2 July 2010, London

Mumbai Port Trust Docks Department (2009) Letter of 13th March 2009 to the Iron, Steel Scrap & Shipbreakers Association of India, the Indian Shipbreakers Association and all ship breakers. www.mumbaiport.gov.in/newsite/otherinfo/circulars/shipbreak130309.htm. Accessed 10 Dec 2009

NGO Shipbreaking Platform (2013) Platform News – green ship recycling in Europe, visit to renovated yard in Denmark. www.shipbreakingplatform.org/platform-news-green-ship-recycling-inEurope. Accessed 26 July 2013

Nijkerk A (2005) Shipbreaking. USA Recycling International, 22 March 2006

Pakistan Observer (2005) Ship-breaking industry gets tax relief. Pakistan Observer, 28 January 2005, Islamabad

Pandey P (2010) High duties hurt Alang prospects. Business Standard (India), 14 February 2010

Paris MoU (2012) Paris MoU Annual Report 2011

Putherchenril TG (2010) From shipbreaking to sustainable ship recycling. Evolution of a legal process. Koninklijke Brill NV, Leiden

Reyes B (2010a) Volatile scrapping rates send mixed signals as market falls. Lloyd's List, 27 April 2010, London

Reyes B (2010b) China claims top spot from India. Lloyd's List, 4 April 2010, London

Reyes B (2012) Protests shut Alang yards after three held over Union Brave blast. Lloyd's List, 16 October 2012, London

Robin des Bois (2007) Global statement 2006 of shipping vessels sent to demolition. Association for the Protection of Man and the Environment. www.robindebois.org/english/sea/global-shipbreaking-2006.html. Accessed 29 July 2010

Robin des Bois (2008) Global Statement 2007 of shipping vessels sent to demolition. Association for the Protection of Man and the Environment. www.robindebois.org/english/sea/global-shipbreaking-2007.html. Accessed 29 July 2010

Robin des Bois (2009) Global Statement 2008 of shipping vessels sent to demolition. Association for the Protection of Man and the Environment. www.robindebois.org/english/sea/global-shipbreaking-2008.html. Accessed 29 July 2010

Robin des Bois (2010a) Global statement 2009 of vessels sent to demolition: the threshold of 1,000 Vessels is reached. Association for the Protection of Man and the Environment. www.robindebois.org/english/sea/global_2009.html. Accessed 29 July 2010

Robin des Bois (2010b) Information and analysis bulletin on ship demolition #18. www.robindesbois.org/english/shipbreaking18.pdf. Accessed 12 Apr 2011

Robin des Bois (2011) Information and analysis bulletin on ship demolition #22. www.robindesbois.org/english/shipbreaking22.pdf. Accessed 12 Apr 2011

Røst OH (2003) Shipbreaking: lives at stake. Metal World, 12 February 2003

Sharma A (2011) Extending the cash buyer's portfolio. In: 6th annual ship recycling conference, 14–15 June 2011, London

Shashikumar VK (2005) Alang the waste coast. Tehelka, 13 August 2005. www.tehlka.com/story_main13.asp?filename=Ne081305alang_the.asp. Accessed 14 Apr 2010

Ship Recyclers' Association of Turkey (2007) Shipbreaking in Turkey. www.gemisander.org/content.php?id=0001&lang=en. Accessed 3 Jan 2010

Ships Monthly (2012) Will SINKEX be sunk? Ships, March 2012. Kelsey Publishing, Cudham

Ships Monthly (2013a) Boldly gone. February 2013. Kelsey Publishing, Cudham

Ships Monthly (2013b) Getting good value for money? October 2013. Kelsey Publishing, Cudham

The Daily Star (2006) 31 July 2006, Bangladesh

The Maritime Executive (2011) MARAD celebrates opening of 1st West Coast ship recycling facility. www.maritime-executive.com/article/2011-02-21-us-maritimeadministration. Accessed 1 Feb 2011

UNESCO (2004). Impacts and challenges of a large coastal industry. Alang-Sosiya Ship-breaking yard Gujarat, India. Coastal Region – Small Island Paper 17

Ul-Hasan A (2011) Development of ship recycling facilities in South Asia: Pakistan. In: 6th annual ship recycling conference, 14–15 June 2011, London

Uytendaal A (2012) How can a beach come up to ISRA standards? In: 7th annual ship recycling conference, 29–30 June 2012, London

Vora R (2010) Handling Ships: from dismantling to building it. International Business Times, 5 May 2010. www.ibtimes.com/articles/22230/20100505/harbouring-ships-from-dismantling-building-it.htm. Accessed 18 July 2010

Wansell S (2011) Market Watch 2011: overview cash buyer volume, prices, forecast. In: 6th annual ship recycling conference, 14–15 June 2011, London

Young Power in Social Action (2007) Shipbreaking in Bangladesh. www.shipbreakingbd.info/Copy%20of%20News.htm. Accessed 20 February 2009

Young Power in Social Action (2010) Shipbreaking in Bangladesh. http://shipbreakingbd.info/story/46%20workers%20killed%20onthe%20shipbreaking%20beaches.html. Accessed 15 Mar 2012

Zakaria NGM (2011) Development of ship recycling facilities in Bangladesh. In: 6th annual ship recycling conference, 14–15 June 2011, London

Chapter 2
The Role and Application of International Law

Before examining in Chap. 3 the national and international legislation that regulates (or should regulate) shipbreaking in the major breaking states, this chapter will briefly consider as background the theory of international law in terms of sources and the mechanics obtaining, including the problems associated with the question of interpretation, enforcement and the settlement of international disputes where they arise. If the new *HKC* enters into force, its effective and successful operation will still depend—at least in part—upon how individual states will interpret and apply the various voluntary guidelines upon which the Convention is based. This is especially relevant to the shipbreaking states and their development of national standards for their shipbreaking facilities, the subsequent inspection of those facilities and the national enforcement of the standards which they set. State sovereignty '*can be used to protect a state against the intervention of international law into its national legal system*'[1] and here, the role of national sovereignty will be significant against international agreement, subject to whatever controls and inducements for compliance and enforcement that relevant legal instruments may include and the ultimate effectiveness of these provisions.

The question of the need for an international law solution to the problems of shipbreaking has arisen from

a) The impact upon both human and environmental health from the accumulated pollution produced by the industry.
b) The growth of international dissatisfaction with the way the industry is and has been operated (although this comes *after* increasingly restrictive legislation has driven the industry from earlier locations).
c) The apparent failure of various voluntary international guidelines (see Chap. 3).

Sands and Peel identify three aspects of the development of international environmental law over the past six decades, namely its reactive rather than

[1] Park (2002). On the other hand, the growth of environmental and human rights law may impose restrictions upon this.

M. Galley, *Shipbreaking: Hazards and Liabilities*, DOI 10.1007/978-3-319-04699-0_2,

proactive approach to events, the catalytic role of science and technology and thirdly the involvement of non-governmental actors and bodies. More recently the growing presentation of international law before international courts has been of note.[2] With regard to the third issue, the involvement of other non-legal centred bodies—and this applies particularly to the shipbreaking industry—the direct actions of various Non-Government Organizations (NGOs) have been very effective in raising public awareness of the hazards involved in the current shipbreaking industry and have prompted a number of decisions from national courts that oppose the direct export (or import) of what are often presented to and by them as contaminated, end-of-life ships. This chapter will therefore also examine the role of NGOs with particular reference to the part that they play in the development of international and municipal law.

The *HKC* is the latest of a wide range of treaties facilitated and developed by the IMO. The chapter therefore concludes with an examination of the IMO's role in the introduction and maintenance of international treaties.

2.1 Sources of Public International Law

Public international law addresses the relationships between states, between a state and its citizens or citizens of another state and, increasingly in the twentieth century, with international organizations. Characterised as '*a system of constant renewal, dynamism and development*,'[3] it has its sources in convention or treaty law, in customary law, identified by the criteria of *opinio juris* and *usus*,[4] and in legal principles, as well as in judicial decisions and the writings of 'the most qualified publicists'.[5] Traditional interpretation of international law recognised no ranking between treaty and customary law but, since the late 1960s, the further development of peremptory norms—*jus cogens*—has risen to pre-eminence[6]

[2] Sands and Peel (2012), p. 23.

[3] Voigt (2008), p. 4.

[4] *Opinio juris*—the expressed opinion of states that their actions have a legal basis (subjective criteria) and *usus*—the actions and official statements of a state, recognised by others as being in accordance with expectations (objective criteria). Van der Driessche (2011).

[5] *Statute of the International Court of Justice*, Article 38 1. '*The Court, whose function is to decide in accordance with international law such disputes as are submitted to it, shall apply: a. international conventions, whether general or particular, establishing rules expressly recognized by the contesting states; b. international custom, as evidence of a general practice accepted as law; c. the general principles of law recognized by civilized nations; d. subject to the provisions of Article 59, judicial decisions and the teachings of the most highly qualified publicists of the various nations, as subsidiary means for the determination of rules of law.*'

[6] Cassese attributes this emergence of *jus cogens* to a combination of the developing countries' fight against colonialism and the efforts of socialist states to formulate peaceful coexistence with states operating different economic and social structures. Cassese (2005), pp. 199–200. The

although a general rule holds that which is the more recent holds priority.[7] Multilateral treaties are considered as new law, binding on the signing Parties, yet treaties that are held to be contrary to accepted *jus cogens* are deemed to be void.

Ranking behind treaty and customary law as sources are general principles of law, which may be employed by courts to fill in gaps left in legal provisions. Although recognised in the UN Charter (Art. 2)[8] and the Statute of the ICJ (Art. 38) and elsewhere,[9] these general principles give guidance on behaviour but, by not being directly applicable, cannot be challenged in international courts.[10]

In matters of the positioning of international and municipal law, a distinction has usually been made between the concept of monism, whereby both are treated as parts of the same order, with international law being considered automatically part of a state's own law, and dualism, where neither system can alter the other.[11] In this latter case, treaties ratified by a state are deemed to be effective only at the international level until ratified into municipal law, at which point they are regarded as municipal law.[12] The path chosen by each state is usually a function of the interpretation of the state's constitution and municipal laws by the state's own courts, whilst the general lack of specific enforcement provisions in international law means that enforcement is often through national courts, national law thereby determining the effectiveness

development of suitable judicial mechanisms to pronounce on determining such peremptory norms and disputes arising was enshrined in the International Court of Justice.

In 2005, the EC Court of First Instance held that *jus cogens* had a high standing in public international law, binding on all individuals and from which derogation was not possible. *Yusuf and Al Barakaat International Foundation v. Council of the European Union and Commission of the European Communities* case T-306/01, 2005. Capaldo (2008), p. 192.

[7] Shaw (2008), p. 123.

[8] Stated principles include the non-use of force, the settlement of international disputes by peaceful means, non-interference in the domestic jurisdiction of other states, equal rights and self-determination of peoples, equal sovereignty of states and the duty of states to co-operate with one another.

[9] For example, the *UN Framework Convention on Climate Change* and the *Outer Space Treaty*.

[10] Lang (1999), p. 159.

[11] The distinction between the two may be reflected in the different doctrines of incorporation and transformation, the former holding that (customary) international law is automatically a part of municipal law, without the need for a formal ratification process, whilst the latter sees the need for international law to be transformed into municipal law by a formal process such as an Act of Parliament before it becomes effective in the municipal sphere. Shaw (2008), pp. 139–140. A third option to monism/dualism may also be offered, namely the monistic theory of maintaining the unity of differing legal systems with each operating in its own sphere, but with the primacy of international law (the Fitzmaurice doctrine). Such a situation may lead to a conflict of obligations when a state acts domestically in a manner that violates its international obligations. Kaczorowska (2008), p. 39.

[12] Kaczorowska (2008), p. 38.

of international actions and legal decisions.[13] The UK, with no written constitution, operates a monism system of common law and a dualist system with regard to statutes, whereby Parliament participates in the conclusion of treaties, whilst the USA, which is also a common law state, runs a dualist system which is written into their constitution, with the President having powers to make treaties, but only with the approval of two-thirds of the Senate.[14] Where a state fails to act in the domestic sphere in accordance with its international obligations, then others may have to take recourse in international provisions for settlement.

2.2 International Law and Sustainable Development

International environmental law represents an evolution in traditional international law in that:

> It puts new limits on State sovereignty; it intrudes into the domestic jurisdiction and territorial integrity of States. It creates greater responsibilities for States and it involves many non-state entities in the process of international law making. The global nature of environmental issues means that national action by itself... may be insufficient, and that significant international co-operation is required.[15]

The development of international environmental law involved a move away from purely boundary-based issues to ones of international or global significance and was accompanied by the need to address these international concerns with a much higher degree of interdependence of obligations than previously, including its integration into the realms of trade and development.[16] The first UN Conference on the Human Environment in 1972 resulted in the formation of the United Nations Environmental Programme (UNEP) and the production of the *Stockholm Declaration*,[17] which called for a co-operative spirit between states for the protection of the environment.

A decade later, in 1982, the UN's General Assembly adopted the (non-binding) *World Charter for Nature*,[18] which encouraged the formulation of national and

[13] Park (2002). In *Australia & New Zealand Banking Group Ltd. et al v. Australia et al* (the *International Tin Council case*) 1990 29 ILM 670 in the UK, the court held that '*the Government may negotiate, conclude, construe, observe, breach, repudiate or terminate a treaty. Parliament may alter laws of the United Kingdom, but the courts must enforce those laws.... Public international law cannot alter the meaning and effect of United Kingdom legislation.*'

[14] US Constitution, Article 2, section 2. The original 1787 Constitution (Article VI, section 2) stated that "...all treaties made or which shall be made under the authority of the United States, shall be the supreme law of the land..." Park (2002).

[15] Park (2002).

[16] Sands and Peel (2012), p. 4.

[17] *Declaration of the Stockholm Conference on the Human Environment* (UNCHE) 1972.

[18] Drafted by the International Union for the Conservation of Nature. Although the question of environmental issues is not specifically addressed in the *UN Charter*, Article 74 of the Charter requires that the policy of the Members '*must be based on the general principle of good*

international legislation to give regard to the protection and safeguarding of nature *vis-a-vis* economic development.[19] The 1987 World Commission's[20] report '*Our Common Future*' also known as the '*Bruntland Report*,' emphasised the need for sustainable development[21] and led in turn to the 1992 UN Conference on Environment and Development (UNCED) at Rio de Janeiro, where the question of international concern over environmental issues was raised to prominence, priorities agreed and the concept of sustainable development endorsed.

The *Rio Declaration*[22] included amongst its Principles the need for precautionary measures (Principle 15),[23] the polluter pays principle (Principle 16)[24] and environmental impact assessments (Principle 17). The 1992 Conference also adopted a number of measures,[25] which included a Programme of Action (Agenda 21), to be monitored by a Commission on Sustainable Development and relating to local involvement in environmental matters which, although adopted at state level, were to be devolved to local authorities for action.[26] Of significance for the marine environment is Chapter 17 of Agenda 21, which established a series of programme for the protection of the oceans and seas, including programmes for integrated management and sustainable development of coastal areas and a programme for marine environmental protection, especially from land-based sources. In this latter

neighbourliness,' taking into account '*the interests and well-being of the rest of the world, in social, economic and commercial matters*.' Sands and Peel (2012), p. 38.

[19] Neither the *Stockholm Declaration* nor the *World Charter* actually defined 'environment' but distinguished between natural resources and ecosystems and man-made living and working environments. Sands and Peel (2012).

[20] The World Commission was formed in 1983 by a General Assembly of the UN, the resolution calling for better ways to address the issues of development and environment.

[21] Whilst international environmental law is an essential aspect of international law of sustainable development, the two are not totally synonymous, economic and social factors also representing elements of the latter. The term 'sustainable development' was given international legal credence by the ICJ in the *Gabčikovo – Nagymaros case*, Sands and Peel (2012), p. 10.

[22] *Declaration on Environment and Development* (the *Rio Declaration*) 1992.

[23] Adoption of the 'precautionary' principle was to be a controversial issue amongst a number of states who regarded its inclusion as a potential barrier to trade and a potential barrier to the development of new technologies. Louka (2006), p. 34.

[24] The 'polluter pays' principle recognised the fact that the effects of pollution are not restricted merely to the producers but may impact upon society in the widest sense. Louka suggests that the internalisation of costs would ultimately be passed on from producer to consumers and is unlikely to receive a positive reception. There has also been demonstrated an international reluctance to pursue the principle with much vigour, witness the reluctance of other states to press for compensation from the Soviet Union for damages caused by the Chernobyl disaster in 1986. Louka (2006), p. 34.

[25] These also included the *Convention of Biological Diversity*, a *Convention on Climate Change*, and a Statement of Principles on Forests.

[26] Birnie et al. cite various views to the effect that Agenda 21 should be regarded more as soft law rather than a legally binding text whose contents governments have been selective in adopting and which does not require ratification, but which nevertheless has effectively underpinned subsequent international agreements. Birnie et al. (2009).

regard, the shipbreaking operations of the sub-continent would appear to be directly relevant. A decade later, a review of progress from the Rio Conference was held at Johannesburg in 2002.

An aspect of the development of international environmental law has been the special consideration given to the less-developed countries, mainly in the southern hemisphere, where poverty and the need for economic development have been regarded as an environmental problem,[27] with economic provisions and the need for special (lower) standards and responsibilities reflecting their ability to contribute to international issues. The conflict between environmental protection and economic development may be a major challenge for many states (see the case of the *Blue Lady*, Chap. 5). Nevertheless, international recognition of these states' sovereignty over their natural resources and national regulations no longer enables them to disregard the consequences of their activities for other states or for common resources. Whilst shipbreaking is basically a sustainable industry, with high levels of recycling and employment, especially in the major shipbreaking states, it is also one that is currently carried out in circumstances which at times appear distinctly unsustainable. Shipbreaking may be essentially a national industry,[28] but there are no boundaries to marine and atmospheric pollution. Both the sea and the atmosphere are now designated common heritage, with Global Commons relating to the marine environment and the Common Heritage of Mankind to the atmosphere, and ships may pass through a number of jurisdictions *en route* to the breakers. The locations of various coastal shipbreaking centres pose potential, if not actual, harm if transboundary pollution is carried to neighbouring states.

2.3 Sovereignty, Compliance and Dispute Resolution

Whilst national sovereignty may traditionally be regarded as a cornerstone of international law,[29] and '*it is within sovereign states that international law is put into practice*,'[30] customary law and specific treaty provisions impose international obligations on individual states. States need to demonstrate good faith in the performance of these obligations and cannot pass or cite their own laws to justify a breach in their failure to meet the needs of treaties.[31] Shaw argues that, with the rapid growth of global communications, together with the increased interdependency of political and commercial life, even the largest of states can no

[27] Birnie et al. (2009), p. 56.

[28] National in the sense that its economic significance and degree of control varies between the practising states.

[29] Cameron (1996), p. 31.

[30] Park (2002).

[31] *Vienna Convention on the Law of Treaties 1969*. Articles 26–27.

longer claim complete sovereignty,[32] yet it is the opinion of more than one interviewee closely associated with the *HKC* that none of its provisions can overrule or by-pass national interpretation once the treaty has been enacted into national legislation. Even the requirement of former owners who seek proof of demolition of a vessel, *e.g.* under facilitated arrangements in China, needs to acknowledge that the presentation of documented scrap disposal is limited to, and a function of, that state's own municipal provisions.

The level of compliance can vary between treaties and between parties to a common treaty. Reasons for compliance are varied and may include: the furtherance of self-interest; the result of promoting states being already in compliance and wanting other states to 'catch up;' the growth of influence of environmental groups etc.[33] Alternatively, non-compliance[34] may emanate from inadequate or vague provisions; an inadequate administration[35]; the fact that the benefits envisaged are deemed not to warrant the costs of compliance; the absence of an appropriate international court etc. The factors are manifold and may vary over time but a state that signs up to a treaty is expected to abide by its provisions.

Compliance may be promoted by a number of measures specifically introduced into treaties for the purpose. On the positive side, direct financial aid to developing nations[36] and the transfer of technical knowledge are now becoming common features of international treaties, designed to overcome a basic lack of capacity, both in terms of resources or administrative structure.[37] On the negative side, the

---

[32] Shaw (2008), p. 129.

[33] Mitchell considers that it is the absence of an international body that can rule on ambiguities that allows states to make subjective and advantageous interpretations whilst still claiming compliance. Mitchell (1996).

Although claiming to be a diligent Party to the *Basel Convention*, India chooses to interpret the presence of asbestos and other hazardous materials on board ships arriving for demolition as being only those that may be represented as *cargo*, all other substances contained within the structure of the ship being disregarded—surely not the intention of the Basel Conference of Parties as a whole—and this has given rise to disputes with other states, but with outcomes usually to the benefit of India. India remains one of the few countries that still operates an active asbestos industry. See Chap. 5—the *Riky* and *Blue Lady* cases.

[34] By non-compliance is meant a failure to meet the legal requirements of a treaty, although some seek to make a distinction between non-compliance and a failure to meet the objectives of a treaty, even though the parties may be in full compliance with the terms of the treaty.

[35] A growing number of open registers appear to offer little or no effective maritime administration, being merely money raising ventures, likely to be operated by others outside the state in question. In addition, the fact that many ships may rarely (or never) actually enter their home ports may drastically reduce the level of flag state control over pollution standards. See also Chap. 4.

[36] Whilst some limited financial provisions were incorporated into the BC, the fact that the Convention was issued only in English required that a significant proportion of that finance had to be devoted to translator services at meetings of the COP—comment from discussions with consultant, 17.1.2011.

[37] As well as treaty provisions, direct payments by some states may be made to assist others in removing problems that may be regarded as having potential and adverse transboundary impacts, e.g. the payments made by the USA, UK and others to Russia for the controlled scrapping of the latter's growing fleet of obsolete nuclear submarines.

threat of international sanctions may be sufficient to convince non-conforming states that the cost of non-compliance may be greater than the benefits obtained (although the cost of sanctions may also impinge upon those states actually employing them). A requirement for regular meetings of Parties and for mandatory reporting of performance to a central body or secretariat is designed to promote compliance by miscreants (although this may not necessarily be effective), but the provision of international bodies with the authority to verify and/or rule on such matters gives compliant states the means to exert a conformity of approach. The 1987 *Montreal Protocol*[38] and its 1990 amendments has been especially effective in this respect, since it was one of the first treaties to include provisions for the transfer of finance and technical support in order to encourage the participation of recalcitrant countries.[39] Compliance is further encouraged via regular meetings of the Parties, annual reporting requirements (and the dissemination of that data) and the formulation of measures which may be applied to non-compliant states.[40] In 2003, Parties to the *Basel Convention* adopted an Implementation and Compliance Committee,[41] offering a '*non-confrontational, transparent, cost-effective and preventive in nature, simple, flexible, non-binding and oriented in the direction of helping parties to implement the provisions of the Basel Convention,*' and paying '*particular attention to the special needs of developing countries and countries with economies in transition,*' the said mechanism being '*without prejudice to the provisions of article 20 on settlement of disputes.*'[42] Enforcement provisions in the *Convention on the International Trade in Endangered Species (CITES)*[43] are

[38] *Protocol to the 1985 Vienna Convention for the Protection of the Ozone Layer.*

[39] The Financial Mechanism (Art. 10) for financial and technical co-operation is operated by an executive committee working with the World Bank and UNEP on behalf of the Parties. Similar financial arrangements were also incorporated into the *Framework Convention on Climate Change.* Sands and Peel (2012), p. 282. On the other hand, whilst financial assistance may encourage participation by states that might otherwise not have the resources (or inclination) to participate, it is accompanied by the risk of promoting free-loaders and those states who demonstrate a reluctance to participate in a process that they might otherwise engage in. Such inducements are also generally more expensive than the application of sanctions (which may only become expensive if they prove to be ineffective). Mitchell (1996), p. 21.

[40] Annual reporting requirements relate to details of each state's production and trade in specified controlled substances both with other Party and non-Party states. Sands and Peel (2012), p. 164.

Also, the non-compliance procedure of the Protocol can be invoked by any of the Parties for investigation by an Implementation Committee, whose report is placed before a full Meeting, which determines the measures necessary to promote compliance—similar (if slightly different) provisions have been incorporated into other subsequent treaties. Similar provisions exist within the 1979 *Geneva Convention on Long Range Transboundary Air Pollution* and to some extent in *the 1998 Aarhus Convention* and the 1997 *Kyoto Protocol.* Birnie et al. (2009), pp. 247–249.

[41] COP6, Decision VI/12, later amended at COP 10, Decision BC-10/11.

[42] Article 20 of the Convention promoted settlement of disputes through negotiation or any other peaceful means of their own choice or failing that through the International Court of Justice or arbitration. Basel Convention (2011).

[43] *Convention on the International Trade in Endangered Species of Wild Fauna and Flora* (CITES) 1973.

deemed to be relatively comprehensive, yet despite numerous resolutions by the Conference of the Parties there remain difficulties in ensuring the proper functioning of the treaty with regard to border controls and the monitoring of international trade in prohibited exports.[44]

The basic premise of a treaty is the observation by all Parties of the obligations contained therein; nevertheless, disputes between states do take place. Although treaties may contain references to dispute settlement via a range of mutually agreed options, various formal measures involving international organisations have also evolved. Aimed principally at the avoidance of war and the maintenance of peace and matters addressing economic, trade and environmental issues etc., the development of international law was associated with the formation after World War I of international bodies such as the League of Nations and the Permanent Court of International Justice (PCIJ). After World War II, the United Nations, the World Trade Organization (WTO) and the United Nations Environment Programme (UNEP) were established to promote and preserve peaceful discourse between states. Since the formulation of the UN Charter, the traditional use of force and blockade etc. as means of resolution have (in the main) been replaced by a non-adversarial approach, centred upon either negotiation and mutual agreement or through the observance of a legally binding decision of a formally authorised third party.[45] The Charter allows disputing states the choice of the means of settlement, including reference to international judicial bodies, such as the UN's International Court of Justice (although this has no specific priority as a forum for settlement).[46] Access to proceedings is limited to states and advisory opinions requested by intergovernmental organisations; access to other international tribunals is less restricted and, although such bodies may have experienced an increasing

[44] Threatened species are listed in the Appendices to CITES, yet states are able to register reservations to specific species, thereby giving them the opportunity to trade in those species with other states with similar reservations or with non-party states, giving rise to problems in global enforcement. Further, the absence of a compensatory scheme to African states for the loss of revenue resulting from the 1989 ban on international trade in ivory may have deterred developing countries from adopting similar conservation measures. Sands and Peel (2012). In a move to raise money for conservation, CITES officials agreed a one-time auction of stockpiled ivory to Japan and China in 2008, but high priced sales by China has prompted the attractiveness of further poaching. Levin (2013). Further, Thailand's legal market in ivory allows illegal African ivory to be 'laundered' and calls for sanctions have been made against eight source, transit and destination countries. Carrington (2013), p. 23.

[45] Article 2.4 of the Charter prohibits the use of force or the threat of force, although this provision was originally introduced to promote and maintain international peace rather than as a means of enforcing international law. Article 33 of the Charter addresses the peaceful settlement of disputes, leaving the choice of option to the parties involved.

[46] Birnie et al. cite Sir Robert Jennings on the ICJ's narrow jurisdiction as reflecting '*a concept of participation in the international legal system that is now seventy-five years old, increasingly anomalous, and out of step with contemporary international society*.' Birnie et al. (2009), p. 251.

workload over the years, such approaches are not necessarily regarded as the optimum road to resolving international environmental disputes.[47]

Negotiations intended to promote consensus and stability may take place directly between the states involved in dispute,[48] with the outcome formulated by the participants, although this may provide the opportunity for a dominant party to exert undue influence. Third parties, which may be states or international institutions, can play an increasingly active role in the diplomatic settlement of disputes. Inquiry involves the establishment of an independent international body to determine the facts of the matter, which may, or may not, be legally binding and which may be the precursor to the subsequent involvement of a judicial body.[49] A third party may further offer its good offices to promote an amicable settlement between the disputing parties, take a more active participation in the process of mediation or, even more actively, take part in the process of conciliation, formally proposing settlement terms, which may or may not be legally binding.[50]

Parties seeking a legally-binding settlement to a dispute may voluntarily turn to arbitration, with the obligation to accept awards in good faith.[51] Certain treaties may include specific, compulsory measures for their 'interpretation and application' by judicial institutions or for judgement from an international court such as the International Tribunal for the Law of the Sea,[52] although such cases are not common. At UNCED and in the development of the *Montreal Protocol* and *Climate Change Convention*, much focus was put upon the need for mechanisms to prevent disputes; the provisions of the *HKC*, on the other hand, leave the choice of method of dispute settlement to the parties involved—see Chap. 6.

Actual recourse to litigation as a means of settling international disputes over environmental law has been somewhat limited (with perhaps the exception of the Law of the Sea). Where a dispute extends across several treaties or across multiple parties, where customary law is somewhat unsettled, or where the jurisdiction of a body is limited—all these may result in a preference for settlement by negotiation rather than litigation, especially if there appears to be a danger of an unsatisfactory

---

[47] Birnie et al. (2009), p. 251.

[48] In some instances, NGOs and even individuals may initiate action.

[49] The inquiry method was an element of the 1899 *Hague Convention for the Peaceful Settlement of Disputes*.

[50] Cassese (2005), pp. 279–280.

[51] Article 37 of the 1907 *Hague Convention on the Pacific Settlement of International Disputes*. Such measures were a characteristic of a number of early environmental treaties, including the 1958 *High Seas Conservation Convention*.

[52] For example, the 1972 *London Dumping Convention* and the 1973/78 *MARPOL Convention*. UNCLOS 1982 allows relevant disputes to be put to the International Tribunal for the Law of the Sea, to the International Court of Justice (ICJ) or to an arbitral tribunal (Art. 281–2), whilst the World Trade Organization scheme calls for a system of compulsory binding arbitration, unless such disputes involve other multilateral environmental agreements, in which case the preference is for such disputes to be settled under the provisions of the relevant agreement (which may have no such provisions for binding settlements). All disputes under the 1991 *Antarctic Protocol* are referred to arbitration for settlement. Birnie et al. (2009), pp. 259–261.

precedent being established.[53] The special chamber established in 1993 by the ICJ to hear environmental cases was abolished in 2006, no cases having been referred.[54]

2.4 The Role of NGOs in Law Making

NGOs are now established participants in the formulation of law at all levels, from international law to national legislation, either as individual organizations, or with other like-minded NGOs, with Intergovernmental Organizations (IGOs), or with other non-governmental organizations.[55] Boyle and Chinkin[56] cite several references which suggest that the participation of NGOs in the formulation of international law has added a level of democratization into the essentially non-democratic process of international law, usually through the acknowledgement of their expertise, particularly in areas such as human rights and environment. Nevertheless, the role of NGOs in the formulation of legislation is essentially catalytic, the actual formulation of law being a matter for governments and their agencies.

A basic strength of NGOs arises from the status accorded to them by the UN's Economic and Social Council (ECOSOC). Article 71 of the UN Charter allows NGOs to enter into consultation agreements with ECOSOC if they are deemed to be '*in conformity with the spirit, purposes and principles of the Charter*,'[57] but UN resolutions allow them also to be invited as observers to attend public sessions of the General Assembly (GA) and Security Council (SC) if they concern matters of relevance to the NGOs.[58] NGO contributions to the development of international

---

[53] For such reasons, no claims were made against the Soviet Union following the Chernobyl disaster, the situation with the UK being complicated by running disputes concerning the pollution of the Irish Sea by nuclear waste from the Windscale plant and the effects of acid rain in Scandinavia.

[54] Closure was deemed to be the result of an apparent lack of difference in costs or procedures from those obtaining in the full court. Further, it was difficult to determine what exactly constituted an environmental case since issues are rarely restricted to a narrow range of issues, but may cover a wide range of aspects of international law. Birnie et al. (2009), p. 255.

[55] The role of NGO pressure was significant in the establishment of the 1972 Stockholm Conference and more than 8,000 NGOs attended the 1992 Rio Conference to network and co-ordinate their policies. Birnie et al. (2009), p. 48.

[56] Boyle and Chinkin (2007), pp. 45 and 66.

[57] ECOSOC Resolution 1296 (XLIV), 23 May 1968.

[58] To be considered for such status, an NGO must have an established headquarters; a constitution that has been defined democratically; a transparent decision-making process and a mechanism for accountability to its members. Funding should primarily be from voluntary contributions. This latter item does not apply to NGOs which are corporate entities or commercial associations such as the International Chamber of Shipping. Such organizations may have objectives that are distinctly at odds with NGOs representing environmental, human rights or social justice issues and which may be acting to secure control and accountability over non-social organisations, often by hard laws, rather than by the self-regulatory approach of commercial bodies. The admittance of NGOs to UN meetings and summits as observers is widespread, but their rights to present papers, to

law may actually be to initiate the process of reform by identifying issues of relevance and raising the level of awareness in the international community, and subsequently by the supply of information and expertise, intensive lobbying of delegates and the production of regular documentation to generate impetus in the formulation of new legislation.

Such a process has been pre-eminent in the development of the *HKC*[59] and prominent amongst the actors in the move to improve the performance of the shipbreaking industry have been a number of NGOs, consisting in the main of environmental and human rights activists. Whilst some NGOs take the role of professional or educational bodies, others are campaigning organizations. An early and vigorous campaigner against traditional shipbreaking practices has been Greenpeace, subsequently joined by the Basel Action Network and later amalgamating with other groups to form the NGO Shipbreaking Platform. Together they have been responsible for the raising of international concern with regard to the impacts of the industry, and instrumental in the development of case law from various national courts relating to the export and import of end-of-life ships (see Chaps. 3 and 5) following extensive lobbying of governments and of the shipping industry. NGOs have been influential in discussions at the IMO both during and after the adoption of the *HKC*; on the other hand, NGOs may be viewed increasingly with suspicion in developing states—the response of India (and others) to adverse reports on their shipbreaking industry has been to designate such sites as restricted areas. Greenpeace in particular is noted for its direct actions, although at times its missionary zeal may go some way beyond what may be deemed appropriate. The Greenpeace campaign against the disposal of the Brent Spar storage unit in the North Sea was ultimately very successful, although based on certain pronouncements eventually acknowledged as inaccurate by the organization.

At national level, NGOs have been successful catalysts, or even direct instruments in helping to develop case law, especially in matters relating to environmental protection, *locus standi* having been granted to broad-based groups such as Friends of the Earth and Greenpeace in acknowledgement of their expertise.[60] The incorporation of various national organizations in the Shipbreaking Platform, *e.g.* the Bangladesh Environmental Lawyers Association and the Bangladesh Institute

speak, or to participate at all depends upon the provisions of the relevant treaty; voting rights are not usually granted.

[59] NGOs have also been especially active and successful in helping facilitate such treaties as *the Torture Convention* (the 1984 *Convention against Torture and Other Cruel, Inhuman or Degrading Treatment or Punishment* and the 2002 Optional Protocol), the *Landmines Convention* (1997 *UN Convention on the Prohibition of the Use, Stockpiling, Production and Transfer of Anti-Personnel Mines and on their Destruction*), and the 1998 *Rome Statute of the International Criminal Court*.

[60] Examples include *R v Inspectorate of Pollution, ex parte Greenpeace Ltd* (*No. 2*) 1994, *R v Swale Borough Council, ex parte Royal Society for the Protection of Birds* 1991 and *R v Secretary of State for the Environment ex parte Royal Society for the Protection of Birds* (*Lappell Bank case*). Greenpeace was also an active supporter of developing countries who effectively initiated and secured the *Basel Ban* at the Second COP in 1994.

of Labour Studies, has been particularly significant in influencing legal developments and court judgments in that country. The presence of local representation is useful also in facilitating local monitoring and reporting.[61]

2.5 The Role of the IMO in International Law

The International Maritime Organization was established at the UN Conference in 1948. Support at the outset was very hesitant, the initial emphasis that the IMO Convention[62] placed upon the need to end discrimination and restrictive practices being viewed by some traditional maritime nations as being to the detriment of open commercial trade,[63] hence the Convention did not finally enter into force until a decade later. Since then, the organization has subsequently grown in strength, support and effectiveness, focussing exclusively on maritime matters and on an international basis.[64] Today membership stands at some 167 states, with many states not formerly associated with the shipping industry—including a number of land-locked states—having come to prominence. The rapid growth of open registers has had a distinct impact upon the composition and governance of the Organization, which was initially the domain of the traditional maritime nations.

Since its inception, the IMO can now number some 55 international conventions[65] and protocols and more than 800 codes. Safety is the dominant principle and the human element is being increasingly addressed. Environment and marine pollution (and subsequently compensation for spillages) were not features of the original Organization, but these issues were integrated into its responsibilities in

---

[61] Birnie et al. point out also that Greenpeace has played a useful role in the monitoring of compliance with treaty requirements, exposing the illegal dumping of nuclear waste in the Barents Sea by Russia, in contravention of the *London Dumping Convention* (Birnie et al. 2009, p. 472), whilst in the *Lappel Bank* case, the RSPB challenged the UK government in the ECJ over their failure to designate an area in the Medway estuary as a Special Protection Area.

[62] *Convention on the International Maritime Organization*, Geneva 1948—title as revised by Resolution A371 (X) of 9 November 1977.

[63] Others considered that the new organization was to operate to the benefit of the (then) dominant countries; a large number of reservations were therefore registered in the earlier years. At its entry into force in 1958 after the formulation convention, just 21 nations, representing 1 million gross tons of shipping, had adopted membership in accordance with Article 74.

[64] In 1975 the original name of the organization was changed from Inter-Government Maritime Consultative Organization, dropping the word ‘Consultative’ to reflect more its decision-making and action-formulating roles, the original Article 2 limiting the IMO to an advisory and consultative role being deleted at the following Assembly in 1977.

[65] In addition, there are a further nine that are still in force but no longer operational as they have been replaced by more recent instruments.

1977.[66] As well as environmental matters, security also featured strongly in the new measures of the 2000s.

The time taken for the development and adoption of IMO treaties can be extensive—the 38 months taken for the *HKC* is considered to be something of an IMO record—and the length of time required for conventions to come into force following adoption has been, and still remains, a matter of concern, 5 years being the average time required. This is not so much a function or result of the IMO itself, but of the speed at which Members actually ratify measures into their own domestic legislation. Agreements generally contain provisions as to the number of states required to enact adoption and a define proportion of the world's fleet that they collectively represent.

In common with much international (and municipal) legislation, maritime legislation usually represents a reactive response to events, yet shipping is an industry in which technical developments can advance quickly; as a consequence, defined legislation can quickly become out of date unless an effective system of update is in operation. By the mid 1950s, the process of introducing even small legislative revisions often took longer to pursue than the original convention to which it referred, many changes requiring a proportion of states higher than that required for the approval of the original IMO Convention. There was, therefore, a danger that hard-won agreements might fall increasingly into obsolescence, that nations might revert back to unilateral action and that the role of the IMO would start to decline before it ever met its potential. Consequently, new provisions for 'tacit agreement' to the technical annexes of conventions adopted since 1972 was introduced, under which it was agreed initially that revisions would come into force automatically unless formal objections were received and the process of the maintenance of legislation was transformed significantly.[67] These changes proved to be so popular that there is now an agreement that revisions to SOLAS (which had been amended numerous times since its adoption) shall not normally be made more frequently than every 4 years.[68]

Despite the plethora of maritime legislation applying to, and accepted by, most of its Members, the IMO itself has no direct powers of enforcement, which is the responsibility of the governments of Member States. Although states may be

[66] Many of the major conventions such as SOLAS and MARPOL are now applicable to the whole of the shipping industry and are widely accepted by the industry. The total annual number of oil spills over 700 tonnes has declined from an average of over 25 in the 1970s to 3.7 in the 2000s, despite the growth in seaborne traffic, the greatest period of improvement being in the 1970s, following the entry into force of MARPOL.

[67] This provision was not included in the *Load Line Convention* and most of its amendments have never entered into force. Churchill and Lowe (1999), p. 272.

[68] *The International Convention for the safety of life at sea (SOLAS)*, 1960 was amended six times after it entered into force in 1965—in 1966, 1967, 1968, 1969, 1971 and 1973. In 1974 a completely new convention was adopted incorporating all these amendments (and other minor changes) and has itself been modified on numerous occasions. International Maritime Organisation (2011a).

reported to the IMO for failure to meet their obligations, there are no penalties or sanctions to follow (unlike the EU). UNCLOS defines the roles and responsibilities of flag, port and coastal states, yet the performance of some states in meeting their obligations—especially with regard to enforcement of shipping standards—leaves much to be desired. In order to promote an effective regime of enforcement, the IMO introduced a Flag State Implementation Committee to facilitate the performance of governments and a system of port state control, whereby port states operate a regime of inspections of visiting ships to compensate for the inadequate performance of various flag states in implementing their own responsibilities with regard to ships on their registers. The subsequent formation of a number of Memorandums of Agreement between Member States in various regions further assists this process by co-ordinating inspections between Members.[69] A voluntary IMO Member state audit scheme also highlights the effectiveness with which Member States implement and enforce IMO conventions with feedback and advice on their performance.

Although the original technical focus of the IMO is today still fundamental, many matters now have political and economic dimensions and the scope of its measures continues to widen to meet issues as they enter into industry and public awareness, yet the IMO does not itself have the authority to adopt treaties, merely to provide and promote consultation and facilitation, and this it appears to do well. The Assembly, which meets once every 2 years, adopts a 6-year Strategic Plan; the current Plan, covering the years 2012–2017, identifies amongst the trends, developments and strategic objectives to be addressed the need for heightened maritime security and safety; heightened environmental consciousness; a shifting emphasis onto people and the role of technology as a major driving force for change as well as enhancing its contribution to sustainable development.[70]

[69] Current MoUs exist for Europe and the North Atlantic (Paris MoU); Asia and the Pacific (Tokyo MoU); Latin America (Acuerdo de Viña del Mar); Caribbean (Caribbean MoU); West and Central Africa (Abuja MoU); the Black Sea region (Black Sea MoU); the Mediterranean (Mediterranean MoU); the Indian Ocean (Indian Ocean MoU) and the Arab States of the Gulf (Riyadh MoU).

[70] International Maritime Organisation (2011b). In addition, the Assembly also adopts a High-level Action Plan, which identifies the actions that need to be carried out by IMO towards achieving the strategic directions, as well as the specific outputs that will be delivered during the next biennium as a result of those actions. See also: International Maritime Organisation, *High level Action Plan of the Organization and Priorities for the 2012–2013 Biennium* [Assembly 27th session, Resolution 1038(27), adopted 30 November, IMO, 2011].

References

Basel Convention (2011) www.basel.int/theconvention/implementationcompliancecommittee/mandat/tabid/2296/default.aspx. Accessed 20 Dec 2012

Birnie P, Boyle A, Redgewell C (2009) International law and the environment, 3rd edn. Oxford University Press, Oxford

Boyle A, Chinkin C (2007) The making of international law. OUP, Oxford

Cameron J (1996) Compliance, citizens and NGOs. In Cameron J, Werksman J, Roderick P (eds) Improving compliance with international environmental law. Earthscan Publications, London

Capaldo GZ (2008) The pillars of global law. Ashgate Publishing Ltd., Aldershot

Carrington D (2013) Call for sanctions against nations blamed for elephant killing. The Guardian, 7 March 2013, p 23, London

Cassese A (2005) International law, 2nd edn. OUP, Oxford

Churchill RR, Lowe AV (1999) The law of the sea, 2nd edn. Manchester University Press as Oxford Road, Manchester

International Maritime Organisation (2011a) Adopting a convention, entry into force, accession, amendment, enforcement, tacit agreement procedure. www.imo.org/About/Conventions/Pages/Home.asp. Accessed 20 Dec 2012

International Maritime Organisation (2011b) High level action plan of the organization and priorities for the 2012–2013 biennium. Assembly 27th session, Resolution 1038(27), adopted 30 November

Kaczorowska A (2008) Public international law, 3rd edn. Routledge, London

Lang W (1999) UN-principles and international environmental law. Max Planck Yearbook United Nations Law 3:157–172

Levin D (2013) From elephants' mouths, an illicit trail to China. The New York Times International Weekly, 2 March 2013, New York

Louka E (2006) International environmental law. Cambridge University Press, Cambridge

Mitchell RB (1996) Compliance theory: an overview. In: Cameron J, Werksman J, Roderick P (eds) Improving compliance with international environmental law. Earthscan Publications, London

Park PD (2002) An Oscar for the environment. Inaugural Professorial Lecture, Southampton Institute, 27 June 2002

Sands P, Peel J (2012) Principles of international environmental law, 3rd edn. Cambridge University Press, Cambridge

Shaw MN (2008) International law, 6th edn. Cambridge University Press, Cambridge

Van der Driessche B (2011) Humanitarian law. Programme. Principles of international law. Diakonia International. www.diakonia.se/sa/node.asp?node=993. Accessed 29 Oct 2012

Voigt AC (2008) The role of general principles in international law and their relationship to treaty law. Nordic J Law Justice 2:12

Chapter 3
Legislation

This chapter will examine the various levels of legal instruments that impact upon the global shipbreaking industry, ranging from international conventions (including IMO-engendered conventions), regional (European) provisions, and selected examples of the national legislation of the major shipbreaking states of India and Bangladesh. The degree of observance of the legal liabilities by national and local state governments varies considerably. However, prior to an examination of these issues, the chapter considers the question of hazardous materials that a ship-for-scrap may contain, since the main contention with current practices of shipbreaking centres upon the types and volumes of such substances and their potential for the damage to human and environmental health consequent upon the regimes of handling, storage and disposal to which they are subjected during the scrapping of ships.

3.1 Hazardous Materials

Hazardous materials are contained within the structure of ships, or carried within spares procured around the world, or found in the residues from former cargoes or in operational wastes. They are significant not only for the potential environmental damage that they represent, but also for the direct risks they present to human health, both to the workers and to the surrounding population, in the absence of adequate safety procedures and protective equipment. A summary of these hazardous substances and their properties is contained in the Appendix. Bans and restrictions on many of these substances are already in place in numerous international treaties and it is the effects of these materials that are the basis of the international campaign to improve the way in which ships are currently demolished and, ultimately, the reasoning behind the new *Hong Kong Convention*.

A wide range of toxic or hazardous substances has traditionally been employed in the construction of ships. Many of these substances such as asbestos, although toxic or hazardous, may remain inert by enclosure during the operational life of the

M. Galley, *Shipbreaking: Hazards and Liabilities*, DOI 10.1007/978-3-319-04699-0_3,

ship; it is only when they are disturbed during the breaking process that their inherent, toxic properties become apparent. Asbestos containing materials (ACM)[1] were widely used in the construction of ships between the 1870s and early 1980s. Asbestos provides heat resistance, electrical insulation and acts as a strengthener when added to other materials. It may still be found in the cladding on fire retardant bulkheads etc., in the glands and gaskets of valves and associated pipework, and in thermal insulation such as the lagging on boilers, steam pipes, etc.[2] Although in recent years some of these substances have been banned and/or replaced by other materials of a more benign nature, they may still be present in large quantities when ships, which were constructed some 20 or 30 years previously, finally arrive at the breakers. The use of fibre glass is prevalent on ships but is not included on the listing of hazardous materials, although over time this material pulverises and may be as dangerous as asbestos.[3]

As well as constituent materials, hazardous substances are released as fumes to the atmosphere from cutting operations (especially from painted surfaces) and as the direct release of gases and solvents from tanks and compartments, and the release of minute particles, *e.g.* of asbestos as it is removed by hand. Liquid substances are released either accidentally or deliberately when tanks and bilges are emptied. Waste oil may be burned on the beaches '*to prevent the sea from being polluted*' (*sic*).[4] Paint, especially the highly toxic, anti-fouling coatings, is deposited on the shore as ships are progressively drawn inland. All these materials find their way to the beaches and the sea and hence to other organisms, particularly humans. They are also spread through the surrounding ecology to the dwellings around the shipbreaking sites by a variety of pathways—by wind, by workers' clothing, via underground aquifers, and by tides.

The report from the first visit by Greenpeace to the Indian shipbreaking yards at Alang and Mumbai in 1999 concluded that:

> ... even if activities were stopped at these sites, the high concentration of TBT in the marine sediment and thus the food chain, will remain in the next 10–20 years. Heavy metals, asbestos dust and poorly degradable pollutants from the combustion processes are also contaminating people living in neighbouring areas. The subsequent damage includes the dangers to health resulting from the further use of ship's materials containing asbestos, and emissions in secondary steel rolling and smelting arising from the paint still remaining.[5]

A follow-up visit to Mumbai made by Greenpeace in 2002 indicated that there had been no improvement in the working conditions at the shipbreaking yards, but an increased level of pollution in the sediment samples taken from the yards, the

[1] ACM is defined as any material that contains more than 1 % asbestos.

[2] Maritime and Coastguard Agency. *Merchant Shipping Notice No. M.1428 1991 Asbestos – health hazards and precautions.*

[3] Blankestijn (2012).

[4] Greenpeace (1999), p. 19.

[5] Greenpeace (1999), p. 20.

result of continued scrapping operations.[6] In addition to these chemical and physical hazards, workers are also subject to biological hazards as represented by vermin and insects that inhabit old ships and to a range of the indirect effects of toxicity from the above substances that may, for example, have been digested by fish and crustaceans in the food chain.

The status of hazardous waste management at the main shipbreaking sites (with the exception of Turkey) is difficult to assess. Little official information from China exists or is available; details relating to the disposal of hazardous waste in India, Bangladesh and Pakistan are also difficult to ascertain since these countries have typically been remiss in reporting their hazardous waste activities to the Basel Convention Secretariat.[7]

3.2 International Legislation

3.2.1 The Basel Convention[8]

Prominent amongst the various international treaties that may (or may not) have an impact upon shipbreaking is the *Basel Convention*, whose proponents and opponents have been diametrically opposed as to its relevance to the shipbreaking industry. The development and adoption of the 1989 *Basel Convention* arose out of international concern in the late 1980s that transboundary shipments of waste were becoming more of a problem of the *dumping* of hazardous materials upon developing countries and to Eastern Europe, these growing shipments in turn arising from the increase in environmental regulation in the producing nations.[9] Although developing countries may actually view such imports as a valuable source of revenue and employment, the fact remains that they represent a growing threat to both human health and the environments of those countries, which also tend not to have the capacity—in terms of infrastructure and financial and regulatory provisions—for their safe handling and disposal. In short, a nation's problematic, polluting wastes are simply exported to countries, which are least able to handle and dispose of them in an adequate and safe manner. Some 160 states are party to the Convention, which entered into force in May 1992. Although the difficulties of

---

[6] Greenpeace (2003), p. 3.

[7] COWI (2009), pp. 47–48.

[8] *The Basel Convention on the Control of Transboundary Movements of Hazardous Wastes and their Disposal adopted by the conference of the plenipotentiaries on 22 March 1989.*

[9] In the late 1980s cases of toxic waste from industrialized countries sent to developing countries resulted in international outrage. Eight thousand drums of chemical waste dumped in Koko Beach, Nigeria, and ships like the *Karin B* sailing from port to port trying to offload their cargoes of hazardous waste made the newspaper headlines. A reinforced international legal framework was called for. European Commission (2007), p. 4.

applying the Convention to end-of-life ships was finally recognised by both the Parties to the Convention and the EU in 2011, an examination of its provisions is important in as much as the subject was the basis of most of the confrontation between the opposing factions.

The Convention was enacted under the United Nation's Environmental Programme (UNEP) and, according to one closely associated with the Convention, was instigated not so much by environmental concerns, and not by shipping issues, but principally by health and safety issues.[10] *Basel* is based upon three legal principles; firstly, the principle of minimisation as to the level and the amount of hazardous waste generated; Article 4 of the Convention imposes a liability upon the producing or exporting state of hazardous waste to develop policies and practices that will minimise these levels. Secondly, the 'proximity principle' requires that wastes are disposed of as close as possible to their point of production.[11] Thirdly, disposal is to be undertaken in a manner based upon '*environmentally sound management*;' '*exporting states are to prohibit the export of waste to The High Contracting Parties where the exporter believes the waste may not be handled in such a sound manner*.'[12] The onus for regulating this trade therefore lies with the exporting countries, rather than importing countries.

Article 6 of the Convention establishes a procedure of Prior Informed Consent (PIC), under which the state of export requires that the generator or exporter of the waste notify, in writing, the competent authority of any importing state of proposed transboundary movements; this shall be responded to, in writing, accepting or refusing the shipment or requiring further information. Written consent shall be the trigger that allows the exporting state to sanction the shipment. Shipments that do not comply with these provisions are deemed to be illegal and criminal, and Parties are to take appropriate measures to prevent and to punish such traffic.[13] The Convention thereby actually goes beyond the basic PIC procedure in as much as it imposes a legal obligation of due diligence on all Parties, requiring them to refuse the export of a ship containing hazardous materials if it suspects that the ship will not be dismantled in an environmentally sound manner.[14] Where it is considered that exported hazardous wastes cannot be correctly handled in accordance with the established contract, then the exporting state is under an obligation to re-import the waste.[15]

[10] Comment from discussion with government adviser on Basel and Hong Kong 17.1.2011.

[11] *Basel Convention* Article 4, 2d.

[12] *Basel Convention* Article 4, 2e.

[13] *Basel Convention* Article 4 General obligations, 4, 5.

[14] Hossain and Islam (2006), p. 38.

[15] Article 8 Duty to re-import. The prospect of allowing the LPG tanker *Margaret Hill* to sail from Southampton to Alang for scrapping in 2009 and subsequently (the potential cost of) having to be recalled to the UK was uppermost in the minds of the UK's Environment Agency personnel involved with the prohibition of the ship's departure, until assurances had been given by the ship's owner that the destination was, in fact, Dubai, and for the purpose of re-equipping the vessel as an

Although shipments between Parties to the convention and non-Parties are prohibited,[16] Article 11 permits the provision of bilateral, multilateral and regional agreements (provided that the environmentally sound management of the waste is observed). This would appear to be an area of weakness whereby economic strength may be used to coerce developing states into agreements in a way similar to that allegedly employed by Japan when pressing for support in their campaign to end the ban on commercial whaling.

The primary point of dispute with regard to the provisions of the Convention relates to the extent to which the Convention does, or does not, apply to end-of-life ships. Opinions on this matter are polarised. From the shipping industry viewpoint, a ship is a ship, especially if it is able to sail under its own power, and by definition cannot therefore be regarded as waste, especially if it represents a valuable asset to another party. According to the proponents of the Convention, a ship can be both a ship and hazardous waste at one and the same time; the distinction as to when it can be proclaimed waste is that point at which it is decided to discard the vessel for scrapping—a point which in practice can be very difficult to determine (at least from an observer's point of view). Comments compiled by the Secretariat of the Basel Convention (SBC) from Parties to the Convention as to when a ship becomes waste cover a wide spectrum, varying from when it has reached the end of its useful life (Mexico), when costs of repair exceed operational profits (Bahrain), or when a contract for its recycling is signed (Japan).[17] Suggestions as to when it might cease to be a ship included when it cannot fulfil the criteria for marine safety and becomes a hazard to the marine environment (Bahrain). In many cases, the proffered definitions are distinctly subjective, but the manifold interpretations illustrate one of the difficulties in applying the Convention to end-of-life ships.

At the second meeting of the Conference of Parties (COP-2) of the *Basel Convention* in 1994, agreement was reached on an immediate ban on the export of hazardous wastes from OECD to non-OECD countries, of hazardous wastes for final disposal, and subsequently, in 1997, on the export of wastes for recovery and recycling (Decision II/12)—the *Basel Ban*. At the following session—COP-3, 1995—Decision III/1 was taken to incorporate the Ban as an amendment to the Convention,[18] constituting a new Article 4A in the convention. The *Basel Ban* was duly incorporated as an amendment in 1995 but is not yet in force, coming into

offshore storage facility. Ref: discussions with Environment Agency personnel, December 2009. See also Chap. 5.

[16] *Basel Convention* Article 9, 5.

[17] Secretariat of the Basel Convention (2004).

[18] Decision III/1 differed somewhat from its predecessor in that instead of reference to OECD/non-OECD countries, the ban would apply to exports of hazardous wastes for final recycling/disposal related to the countries listed in Annex VII of the Convention (Basel Convention Parties that are Members of the EU, OECD and Liechtenstein) to non-Annex VII countries (*i.e.* all other parties to the convention).

effect when adopted by 68 of the original 90 Parties to the Convention.[19] However, in 1997 the European Union incorporated the provisions of the ban into its *WSR*, and hence made it legally binding on all Member States. The subsequent acceptance by both the EU and by many Parties to the *Basel Convention* that the Convention cannot effectively be applied to end-of-life-ships—see Chap. 6—may lead to a more rapid acceptance of the Amendment.

The primary questions of the definition of waste in terms of the point at which a ship might be deemed to be waste, and the matter of whom therefore might be considered the exporter of such a vessel in relation to the *Basel Convention* and European legislation, were examined on behalf of the Norwegian Ministry of Environment by Prof. Ulfstein,[20] whose conclusions were subsequently denounced by proponents of the Convention. Waste is defined by the Convention as;

> …substances or objects which are disposed of or are intended to be disposed of or are required to be disposed of by the provisions of national law[21]

and

> Disposal means any operation specified in Annex IV to this convention[22]

Categories of wastes to be controlled are those listed in Annex 1, unless they do not possess any of the characteristics contained in Annex III. Although ships are not specifically included within the Convention (the same might also be said for any number of other products), it is the hazardous substances contained within those products that are specified within the Convention. At the Seventh Conference of the Parties in 2004, a decision[23] was taken to note that a ship may be both waste (as per Article 2 of *Basel*) whilst at the same time still being defined as a ship under other international rules. Further, it was recognised that many ships contain hazardous materials which may become hazardous wastes according to the annexes to the Convention. The debate arising relates to exactly *when* such a definition becomes applicable. Since Annex IV contains listings of disposal options which have and do not have the possibility of recovery or recycling, and Article 4 refers to both, then the ban relates both to export for final disposal and for recovery/recycling (or to a combination of the two).

'Intention' is a concept that is difficult to define, both from the point of view of time frame—the decision to dispose of a vessel may be taken some considerable time before its trading operations are concluded—and from the difficulty of other parties to establish exactly when such a decision has been made. It is, therefore, a process that is very easy to conceal. In his report to the Norwegian Ministry of Environment, Ulfstein considered that no distinction had been defined as to whether

[19] Agreement reached at COP-10, Cartagena, Colombia, 21 October 2011. The amendment has already been ratified by some 51 Members.

[20] Ulfstein (1999).

[21] *Basel Convention* Article 2 (1).

[22] *Basel Convention* Article 2 (4).

[23] Decision VII/26 COP 7 October 2004.

such waste can still be considered a ship under international law or whether a ship deemed to be for disposal is still to be regarded as waste if it carries cargo en route to its disposal site, adding irrespective of whether or not a ship was still considered to be such, once the decision to scrap had been made, that rendered the vessel waste, although the decision to scrap at a more distant time may add uncertainty to this actual definition.[24]

A consideration of Ulfstein's comments is relevant in that they highlight the difficulties in identifying not only the disposal decision point, but also the export and import states and the liabilities of flag and port states; it is the distinct nature of ships that they can exist in a dual capacity that sets them apart from other waste matters. Applying the label of waste to a ship for disposal is fraught with difficulties and perhaps with contradictions. If it is decided that a ship shall be sent for scrap at some future date and meanwhile shall reside in lay-up and out of commission, then the definition of waste and the decision to dispose according to *Basel* appears to be relatively straightforward. On the other hand, if it is decided that a ship shall be sent for scrap in a year or more and in the meantime shall continue in trade, then the question becomes very much less certain—a cut-off limit is not defined. Perhaps there is a call for new terminology here; since the vast majority of ships arrive at the breakers under their own power and in ballast, in that sense it is contended that, although designated waste, such a vessel may still retain the definition of ship but may, perhaps, be regarded as *mobile waste*, a title that may appear somewhat flippant, but does recognise the aspects of a ship being both a ship and waste in law and simultaneously. Alternatively, sailing to a nearby port with cargo for discharge prior to a final departure for scrapping is a common occurrence that offsets some of the costs associated with delivery on behalf of the owners and makes both economic and environmental sense. In the latter instance, the vessel is still in commercial operation and presumably in certificate, with both the ship and the cargo insured. As such, should the waste classification be appropriate only at the final leg of the voyage to the breakers after discharging cargo? Perhaps a more appropriate designation prior to its final discharge might simply be *intended waste*. This scenario certainly adds to the difficulty of determining who exactly is to be regarded as the exporter. If a vessel deemed to be fully in commission discharges cargo in India prior to a final voyage to Alang, then in practice, the states of export and import are one and the same, in essence there might seem to be no State of Export. The practicalities of the shipping industry make decisions on such categorisation of wastes which have other distinct primary functions infinitely more complex that those wastes which are simply inert materials.

The decision to dispose of an article is central to both the Convention and to European waste law. The point of decision to dispose of a ship (for scrapping) may be indicated by any number of factors, including the completion of a contract with a breaker or taking a ship out of service, out of certificate or out of registration, etc. Some of these actions may apply equally to a vessel to be disposed of for future

[24] Ulfstein (1999), p. 7.

trading under a different owner as to a disposal for scrapping. The problem of defining the point of a disposal decision is further complicated by the ease with which such a decision may be concealed, including transfer of ownership whilst a vessel is at sea and hence beyond the jurisdiction of any state; this is not uncommon. It is also complicated by multiple changes of ownership over a relatively short period, a common practice which may be enacted to obscure the actual ownership (and hence liability), often facilitated by registering the ship under certain open registers which promote and assist this process of anonymity—see Chap. 4. An alternative is for the owner to sell the vessel to a state that is a non-Party to the Convention. It is the opinion of the government adviser interviewed that this question of the point of decision to discard is, however, considered to be of less significance than the point at which that decision is exercised.[25]

Ulfstein further considered the question of which state is the state of export of a ship departing for the breakers. Article 2 (10) of the convention defines the state of export as that from which the physical movement '*is planned to be initiated or is initiated.*' In choosing between the two options, Ulfstein dismisses the former, since the place at which decisions are made may be impossible to define and other Articles of the Convention add weight to the case of the state from which the actual movement of the 'waste' actually begins, Article 8 referring to the state to which hazardous wastes must be returned if the movement cannot be undertaken in accordance with the contract. From this, he concludes that if the state in which a vessel becomes waste is not the flag state of the vessel,[26] then the flag state does not have to ensure permission to the shipment of waste from an importing state. This in turn gives rise to the problem that a port state has no jurisdiction to demand that PIC exists before a foreign vessel sails for scrapping and no jurisdiction at all once a vessel has sailed. If, however, ships are determined to be ready for scrap by their owners when on the high seas, there will be no exporting state at all and the provisions of the Convention will have been circumvented.[27] He concludes therefore that an owner may avoid the requirements of *Basel* by arranging for his vessel to become waste either in a foreign port or on the high seas; such avoidance possibilities might be overcome by amendments to the Convention along the lines of the *London Dumping Convention* (see below), which places obligation fully upon flag states.[28] However, if significant amendments were to be made then perhaps they should follow the lines of defining quite clearly the position of end-of-life ships within the Convention.

The question of ships for disposal and the *Basel Convention* was first debated by the Basel Parties in January 2002[29] and the question of the relevance of the

---

[25] Discussions held 17.1.2011.

[26] Indeed many ships, especially those registered with open registries, may rarely or never actually visit their port of registration.

[27] Ulfstein (1999), pp. 11–13.

[28] Ulfstein (1999), p. 13.

[29] Fourth Session of the Legal Working Group of the *Basel Convention*, 10 January 2002.

Convention to the subject was generally agreed. In a paper jointly written for the Legal Working Group,[30] BAN and Greenpeace hold that the Convention goes far beyond a basic prior informed consent routine (Article 6) and refute the earlier conclusion of Ulfstein that port states have no jurisdiction over vessels destined for scrapping with regard to scrapping in other states or that flag states are not responsible for the activities of vessels under their flag—this has now been demonstrated by a number of decisions by various national courts who have laid restrictions on vessels bound for scrap, in some cases citing the provisions of the *Basel Convention*.[31]

The paper asserts that the biggest error in the Ulfstein report is the way in which it ignores the obligation of Parties beyond the Convention's PIC provisions, irrespective of their status—*i.e.* whether state of export, import, transit, flag or port state—to prohibit the export of hazardous wastes where no formal written consent exists or where it is believed that the wastes will not be handled in an environmentally sound manner, and '*it must be noted that most shipbreaking operations around the world at this time* [*2002*] *do not meet this criterion*.'[32] It disputes the conclusion that port states have no jurisdiction over foreign vessels with regard to environmental protection, since the obligation of flag states (which are Party to the Convention) to ensure the existence of consent from the importing state is contained within Article 4 (7) (a), whilst Article 94 (1) of the *United Nations Convention on the Law of the Sea* (*UNCLOS*) requires all states to exercise its jurisdiction and control in administrative, technical and social matters over ships flying its flag.[33] The paper does, however, agree with Ulfstein's assumption that the vessel itself, and not only the individual hazardous substances that might be contained aboard, should be considered hazardous waste under Article 1 (1).

The BAN/Greenpeace paper also takes exception to the shipping industry's claim that a ship cannot be both a ship and hazardous waste at the same time whilst it can sail under its own power. The greater part of the shipping industry and the shipbreaking states are adamant that the Convention applies only to the transboundary shipment of wastes and not to ships which, until demolished, remain intrinsically ships and not wastes. The Convention makes no reference to whether the waste is capable of operating under its own power; it does not exclude wastes that may have subsequent economic use and, apart from the single exception of PCB, defines no concentration level at which a waste may be deemed hazardous.[34] India's Expert Committee appointed by the Supreme Court determined that approximately one percent of a ship's physical constituents comprise hazardous waste as per the definitions of the convention.

---

[30] Basel Action Network and Greenpeace International (2002), p. 1.

[31] See Chap. 5 for examples.

[32] Basel Action Network and Greenpeace International (2002), p. 4.

[33] Basel Action Network and Greenpeace International (2002), p. 6.

[34] Basel Action Network and Greenpeace International (2002), p. 7.

On the question of *Basel*, both the (majority of) owners of ships-for-disposal and of shipbreaking sites would appear to be in agreement on the non-applicability of the Convention, since both parties gain economic benefit from the absence of formal pre-cleaning. In a presentation that seeks to demonstrate the bias of the Convention towards the developed and waste producing nations, Shrinivasan[35] proclaims that the Convention shows a complete disregard for the economics of both the free market and developing countries, taking as it does '*the high moral ground*' on matters of environment and public health. He goes on to condemn the ability of groups such as Greenpeace—groups with no accountability—to exert influence over the development of international law that have a major impact upon both developing and least developed countries.

This would appear to be inaccurate on both counts; firstly, *Basel* seeks a ban on the uncontrolled export of hazardous materials, a problem that can be addressed by a combination of pre-cleaning or the provision of facilities and procedures at the receiving sites that would ameliorate the impacts on human and environmental health. Secondly, although Greenpeace's campaigning abilities have been well developed, their ability to directly affect a wide international forum[36] of national states without some background of moral argument would appear to be crediting them with more than their intrinsic powers. The joining of forces with other international environmental NGOs, including organizations from some of the shipbreaking nations, does add a growing credence.

Shrinivasan's arguments are to some extent contradictory in his claim that to ban access to (hazardous) recyclable waste restricts access to such waste by developing nations, whilst increasing access in developed countries, which by their own regulations render domestic recycling 'unviable.'[37] He cites as an example the US Reserve Fleet, elements of which have been lying in reserve anchorages in the US in some instances since World War II. In fact the US Government now has to pay high prices to subsidise the few remaining US breakers to scrap the ships at home; this action is the result of a combination of US legislation against the export of highly toxic ships overseas and the long period of time, and hence of deterioration, that has affected these ships.[38] Waste he defines as material having no further economic value; in this instance it represents a significant resource for the rolling mills dependent upon the shipbreaking yards. This is not in accord, certainly, with numerous judgements of the ECJ, which hold that wastes are wastes up to the

[35] Shrinivasan (2001), p. 15.

[36] The ability of Greenpeace to raise both public and international resentment against the oil company Shell on the subject of the Brent Spar storage platform in 1995 was a resounding success, even if some of their assertions ultimately proved to be inaccurate.

[37] Shrinivasan (2001), p. 1.

[38] In some instances the poor performance of US breakers in dismantling units of this fleet has been so inadequate that the government agency MARAD responsible for the supervision of these ships has been obliged to recall them from the appointed breakers. The export of four naval auxiliaries for breaking in the UK in 2003 (the Ghost Ships) caused extensive lobbying on both sides of the Atlantic and a ban on the export of further ships for breaking overseas.

point of their re-use. This, interestingly, would therefore more than adequately classify the ships of the Reserve Fleet as waste. His assertions that *Basel* is discriminatory, in contradiction of free trade principles, the spirit of the General Agreement on Tariffs and Trade (GATT) and the World Trade Organisation (WTO), and in contravention of sovereign rights, would appear to place the cost of such freedoms squarely on the shoulders of those directly involved in the breaking activities. *Basel* does not basically seek to eliminate transboundary shipments in total, but to regulate them in terms of the impacts that they currently engender in developing nations.

Andersen concludes that, in its present form, the Convention does not have sufficient potential for improvement, given that it was not developed with the subject of shipbreaking in mind, and hence is not fully compatible with the special characteristics of the shipping industry. Given the importance of shipbreaking to the economies in which it is practised, he suggests that rather than move the industry back to economies best equipped (technologically) to deal with the hazards involved, a better solution is to developed the existing sites and the procedures that they operate.[39] This is basically in keeping with the provisions of the *HKC*, but does not recognise the option of removing at least a part of the hazardous content of a ship before it is finally despatched.

There would appear to be much validity behind the provisions of the *Basel Convention* and its application to end-of-life ships in that it defines a responsibility to prohibit the transboundary movements of hazardous wastes to countries deemed less capable of handling them in a manner that equates to sound environmental management. Yet the very nature of ships that allows them to be both wastes and operational vessels which might be carrying cargo, and the ease with which ownership and the intentions of owners may be changed, renders categorisation difficult to be made with any great certainty and leaves practical and important gaps in the application of the convention.[40]

It is the recommendation by BAN, a fervent supporter of the application of the *Basel Convention* to shipbreaking, that a pragmatic and timely approach be developed to the problem of both intent and spirit of the Convention, which:

> ...from the outset was born to counter the human rights abuses and environmental injustice engendered when toxic wastes are freely traded without restraint in the global market place.[41]

It is further held by BAN in their subsequent opposition to the *HKC* that the new treaty does not address these basic issues. Meanwhile, few ocean-going ships sent for breaking are subject to the PIC routine.[42]

In summation, the *Basel Convention* would appear to be (and is increasingly held to be by various parties and states) applicable to end-of-life ships, yet particularly

[39] Andersen (2001), p. 6.

[40] Bhattacharjee (2009), p. 197.

[41] Basel Action Network (2005), p. 1.

[42] Bhattacharjee (2009), p. 213.

difficult to apply and enforce, since the point of decision of scrapping is difficult to determine and easy to conceal or avoid. Ultimately, an acknowledgement of these difficulties was made by the European Commission and by the Parties to the Basel Convention in 2011. A subsequent report issued by the European Maritime Safety Agency reaffirmed the opinion that the export ban on ships under the *WSR* was hardly enforceable due to the difficulty of identifying the decision to dispose and the lack of recycling capacity in the OECD countries.[43] To this explanation could be added the ease with which the sale and/or reflagging of a ship can (and often does) take place prior to a ship departing for the breakers. The resultant *HKC* shares the onus of liability between both exporting and importing states.

In December 2004, Parties to the Convention adopted a set of technical guidelines on ship dismantling. These guidelines[44] were voluntary in nature and intended to advise the shipbreaking nations on how to improve the environmental performance of their shipbreaking facilities—see Sect. 3.5 below.

3.2.2 United Nations Convention on the Law of the Sea (UNCLOS) 1982

UNCLOS is a comprehensive regime of law relating to the sea and is based on the interconnected relationship of all issues and problems relating thereto. It encompasses issues of state jurisdiction over maritime areas; economic activities; the protection and preservation of the marine environment; marine science and technology; the roles and responsibilities of flag, port and coastal states (and a state may be all three simultaneously) and the resolution of disputes arising. Parties have a general duty to protect and preserve the marine environment.[45] States are also obligated to ensure that activities carried out under their jurisdiction do not cause damage or pollution to other states and their environment[46] and do not transfer damage or hazards from one area to another or transform one type of pollution into another[47]; to monitor the activities which they permit to determine whether they are likely to pollute the marine environment[48]; to assess the potential effects of planned activities[49] and to prevent, reduce and control pollution of the marine environment by dumping[50] and from land-based sources.[51]

[43] European Maritime Safety Agency (2011).

[44] Secretariat of the Basel Convention (2003).

[45] UNCLOS Article 192.

[46] *UNCLOS* Article 194.

[47] *UNCLOS* Article 195.

[48] *UNCLOS* Article 204.

[49] *UNCLOS* Article 206.

[50] *UNCLOS* Article 210 and 216.

[51] *UNCLOS* Article 207 and 213.

In terms of scope, *UNCLOS* does not cover shipbreaking *per se*. It does, however, recognise the role of flag states, who are obliged to enforce the international rules and standards, and exercise jurisdiction in control in administrative, technical and social matters over the ship,[52] including their own national pollution laws and regulations in the face of any violations.[53] A ship on an international voyage *en route* to a breaker's should comply with all applicable international rules and standards, and that is the responsibility of the flag state.[54]

Port states have discretion for enforcement in their territorial waters, and the right to take measures to prevent a ship from sailing, which does not meet the international rules and standards governing seaworthiness and thereby threatens damage to the marine environment.[55] To this end, many states have adopted a Memorandum of Understanding (MoU) on Port State Control to ensure a uniform approach to port state enforcement. The coastal state also has enforcement rights over ships committing discharge violations against relevant international rules or standards with regard to the prevention, reduction, and control of pollution of the marine environment of its territorial sea or EEZ.[56] It is also argued that the transboundary movement of end-of-life ships containing hazardous materials is prohibited under Article 19.

Under the provisions of *UNCLOS*, the term 'waste' is only used in the context of dumping at sea and not the point at which a ship becomes waste for the purposes of disposal at sea. However, under the *London Dumping Guidelines*, it is suggested that a ship would become waste when a permit for that operation was granted. Similarly, *UNCLOS* does not define when a ship ceases to be a ship, but since the Convention only addresses two contexts, including the rights and duties of states with regard to ships which are operational and engaged mainly in an international voyage for commercial purposes,[57] a ship might be construed to cease being a ship once it ceased to operate as such (in the case of *UNCLOS*—the point of disposal at sea). This is also the point at which a ship may cease to be registered as such. A ship that is towed to its place of disposal thereby has already ceased to operate as a ship.[58] In this regard, the provisions of *UNCLOS* may be taken as similar to the provisions of the *Base Convention*.

When a ship becomes or is found abandoned or scuttled on land or at sea without a permit, this becomes a violation under *UNCLOS*[59] and the *London Convention* (see below) for which the flag state is likely to be held responsible. However, since a number of ships which may be so abandoned have their ownership deliberately

[52] *UNCLOS* Article 211 and 217.

[53] Hossain and Islam (2006), p. 39.

[54] United Nations (2002).

[55] *UNCLOS* Article 219.

[56] *UNCLOS* Article 220.

[57] The second context relates to the requirements which must be met prior to dumping a ship at sea.

[58] United Nations (2002), p. 14.

[59] *UNCLOS* Article 210 and 216.

hidden through various open registers—see Chap. 4, then the enforcement of such a responsibility often becomes difficult if not directly impossible.

3.2.3 The Convention on the Prevention of Marine Pollution by Dumping of Wastes and Other Matter (the London Dumping Convention) 1972 and Its Protocol 1996

The *London Dumping Convention* was the first global convention designed to protect the marine environment from human activities. Following on from the *Oslo Convention for the North-East Atlantic*, it requires Parties to take effective measures to prevent marine pollution by disposal of wastes into the sea, the basic obligation being in relation to the deliberate disposal of waste from vessels at sea. Issued in 1972, the Convention entered into force in 1975 and called upon contracting Parties '*to take all practical steps to prevent the pollution of the sea by dumping of wastes and other matter*' including '*any deliberate disposal at sea of vessels... platforms or other man-made structures.*'[60] Dumping of wastes and materials listed in Annex I to the convention is prohibited. Dumping of waste listed in Annex II may be permitted, subject to the granting of a prior special permit, or to the granting of a prior general permit in all other cases.

In 1996, the Convention was updated and replaced by the 'London Protocol,' itself based on the precautionary and polluter pays principles, and covers the same range of structures as defined in the original Convention, but adds '*any abandonment or toppling at site of platforms or other man-made structures for the sole purpose of deliberate disposal*,' an amendment inserted to reflect the provisions of Article 60 of *UNCLOS* 1982.[61] A significant difference from the 1972 Convention lies in the inclusion of a 'reverse listing approach,' under which the dumping of all materials is prohibited unless they are listed in its annex. In principle, both the 1972 Convention and the Protocol allow the disposal of ships at sea,[62] but the Protocol's Specific Guidelines for Assessment of Vessels, adopted in 2000 detail the measures necessary to decontaminate a vessel prior to such disposal, measures which might

[60] *London Dumping Convention* 1972 Article III (1)(a)(ii).

[61] *UNCLOS III* 1982 Article 60.3 Artificial islands, installations and structures in the exclusive economic zone.

[62] For vessels to be disposed of at sea, Parties to the *London Dumping Convention* adopted Specific Guidelines for Assessment of Vessels in 2003. Chapter 4 of the Guidelines lists the potential sources of pollution that should be addressed and items of vessels that potentially contain substances of concern. The Guidelines recommend the removal of materials remaining in tanks, piping or holds of vessels and the removal of all drummed, tanked, or canned liquids or gaseous materials. For ships that are sunk as diving sites (as distinct from artificial reefs), care is taken to remove all protuberances such as masts etc., which might form a snagging hazard for divers, and large holes are cut into the hull and superstructure to facilitate easy access and exit.

apply equally to ships to be scrapped.[63] The deliberate sinking of ships—especially naval vessels and government-owned ships—under the old maritime customary law of abandonment, whilst still relatively infrequent, seems to be a popular option in the USA, where the limited shipbreaking capacity requires that the disposal of ships by scrapping can represent a high cost, rather than revenue, to the authorities.

Although the cost of the necessary pre-cleaning prior to sinking may also be high, it would still appear to represent a cheaper option than scrapping under an environmentally safe management regime. However, disposal of wastes at sea is not viewed as an isolated waste management option, but requires an assessment of all alternative waste disposal options before permitting the dumping of wastes at sea.[64] The granting of approval also requires the production of an impact hypothesis—a statement of the expected consequences of disposal[65]—and a pollution prevention plan,[66] which incorporates best environmental practices. Like the *Basel Convention*, the question of shipbreaking was not discussed when the Convention was being compiled, hence its application to the industry may be debated. Unlike *Basel*, the *London Convention* does not restrict wastes to hazardous materials.

3.2.4 The International Convention for the Prevention of Pollution from Ships (MARPOL) 1973/1978

MARPOL is the main convention covering the pollution of the marine environment by discharges from ships, whether from operational or accidental causes. This Convention, modified by the 1978 *Protocol* (*MARPOL* 73/78), was developed from the 1954 *International Convention for the Prevention of Pollution of the Sea by Oil* (*OILPOL*), following the stranding and breaking up of the tanker *Torrey Canyon* in 1967, and the resultant, extensive pollution of the English Channel. Nevertheless, whilst this accident was a major catastrophe, *MARPOL* recognised that, overall, day-to-day operational pollution was a greater threat.

[63] See IMO 2003 Guidelines on ship recycling Assembly 23rd session. Resolution A.962 (23) adopted 5.12.2003, Item 9.7, IMO, London.

[64] United Nations (2002), p. 12.

Annex 2 of the Protocol defines a hierarchy of waste management options as (1) reuse; (2) - off-site recycling; (3) destruction of hazardous constituents; (4) treatment to reduce or remove the hazardous constituents; (5) disposal on land, into the air, and in water. A comparative risk assessment is required for each of these methods (Art. 3.3) examining the potential environmental (Art. 3.3.1), and health impacts (Art. 3.3.2), technical and practical feasibility (Art. 3.3.3) and economic factors (Art. 3.3.4).

[65] *London Convention* Article 7.1.

[66] *London Convention* Article 4.1.

The Convention is significant in that it provided a catalyst to (then) standard shipbreaking practices through its scheduled phase-out of single-hulled tankers. Amendments to Regulation 13G of Annex 1 brought the phase-out date for Category 1 tankers to 2005 (previously 2007) and for Category 2 and 3 tankers[67] to 2010 (previously 2015).

Its relevance to the industry also lies with the discharges from ships sent for breaking, and which take place within the territorial waters of the shipbreaking state. Any oily wastes found on board must remain on board for subsequent discharge to shore reception facilities and not discharged to the sea, states being obliged to provide such facilities.[68] Annexes to the Convention require that appropriate waste reception facilities be provided for ship-generated wastes at oil loading terminals, repair ports and other ports,[69] and to all ports having ship repair or tank cleaning facilities.[70] In addition, reception facilities for ozone depleting substances and equipment are to be provided at ship recycling yards.[71] Since a vessel arriving for scrapping is usually driven ashore under her own power (and hence must be in compliance with valid operational certificates up to that point), preparatory removal of the named substances prior to beaching—both to decontaminate and lighten the vessels—must be governed by the Convention. The applicability of this Convention might appear to be an argument difficult to refute for those who argue that a ship cannot be other than a ship until demolition begins. However, shore based facilities for such decontamination would appear to be very much absent at certain sites, and the Correspondence Group at the MEPC in January 2001 questioned whether, once a ship is beached, *MARPOL* 73/78 remained applicable.[72]

MARPOL is one of the more important international maritime treaties and came into force with two Annexes covering oil and liquid substances in 1983. Over the next 17 years, a further four Annexes were added, covering other forms of pollution.

[67] Category 1 tankers are also know as pre-*MARPOL* tankers, Category 2 and 3 tankers as *MARPOL* tankers and smaller tankers.

[68] Developed by the MEPC, the IMO's Port Reception Facility Database (PRFD) as a module of the IMO Global Integrated Shipping Information System (GISIS) became a live on the internet on 1 March 2006. Ref: International Maritime Organization (2006)

[69] *MARPOL* Annex I Regulation 12(1).

[70] *MARPOL* Annex I Regulation 12(2)(c).

[71] *MARPOL* Annex VI Regulation 17(1)(c).

[72] International Maritime Organization (2001).

3.2.5 *The Convention on Control of Harmful Anti-fouling Systems for Ships 2001*

The growth of barnacles, molluscs and algae on the hulls of ships decreases the speed of ships through the water and hence adds to operating costs. To combat these problems, various anti-fouling compounds have been applied to hulls, one of the more effective being anti-fouling paints containing toxic organotins such as tributyl tin (TBT). Such substances are highly toxic and persistent in the marine environment, and may even enter the human food chain. As ships are dragged further up the beach during demolition, anti-fouling paints are scraped from hulls and deposited on the beaches, where they are then mixed with the soil and water and thereby spread to the coastal environment. Toxic fumes are released into the atmosphere during torch cutting operations and also whenever the recovered steel is fed into furnaces.[73]

The *International Convention on Control of Harmful Anti-fouling Systems for Ships* sought a total ban by 2008 on the use of such substances on the hulls of ships flying the flag or using the ports of participating Parties. Existing coatings of TBT paint were to be coated over with a sealer, which made it difficult to remove the bottom layer of TBT paint. For this reason, sand blasting to remove TBT paints prior to a ship being exported for scrapping might be considered a mandatory requirement.[74]

3.2.6 *International Convention for the Control and Management of Ships' Ballast Water and Sediments 2004*

Although much attention is directed towards the more obvious hazards of ship-breaking, such as asbestos and PCBs, the question of ballast water discharge was until recently somewhat less acknowledged, despite the fact that the dangers that it represents can have huge environmental impacts on the marine areas concerned and is one of the ecological impacts of shipping that has proved difficult to counter.

When cargo is discharged, ships take on large quantities of ballast water to ensure stability and safety of the vessel; on arrival at a destination for loading, such ballast water is discharged, the water often containing many organisms from the location at which it was taken on board. In favourable conditions, such invasive species may multiply rapidly, to the detriment of indigenous species and possibly to the detriment of human health; such contaminations may be irreversible.

[73] Putherchenril (2010), p. 130.

[74] *International Convention on the Control of Harmful Anti-fouling Systems on Ships*. Adopted: 5 October 2001; Entry into force: 17 September 2008.

The modest size of the world's merchant fleet in the pre-WWII period meant that the problem was not significant; however the rapid growth in the size of ships multiplied the extent of the problem, with large crude oil tankers accounting for much of the total discharge.[75] Despite the inclusion of various provisions in other legal instruments,[76] it was determined that a specific international instrument was required, and in 2004, the IMO adopted the *International Convention for the Control and Management of Ships' Ballast Water and Sediment*, which by October 2013 was not yet in force.[77] Included within the Convention are restrictions as to where ships might safely exchange their ballast waters whilst under way, at specified minimum distances from the nearest land and in waters of minimum specified depth. Numerous methods have been devised for the on-board treatment of ballast waters[78] and many options are presented to meetings of the IMO's Marine Environmental Protection Committee for approval.[79] It is likely that it will take several decades for the world's fleet to be equipped with adequate cleaning systems following the adoption of the new convention.[80]

Ships currently *en route* to their scrapping destinations take on board ballast water for stability purposes, which has to be discharged prior to actual beaching to render the vessel as light as possible, but the provisions of the Convention, which is not yet in force, appear to be largely ignored by the shipowners and breakers, leaving the coastal waters adjacent to the breaking sites vulnerable to invasive species.

[75] Crude oil tankers often make return voyages for loading under ballast, whilst bulk carriers may sail in triangular routes, thereby reducing the need for voyages under full ballast. Container ships operate mostly in liner trades, with full loads in both directions. The estimated volume of ballast water is set at some 3 billion tonnes per year, although no reliable, official statistics are kept. Wijnolst (2010), p. 6.

[76] For example, the obligations as defined by Articles 192, 194(1)(5), 195(2), and 196(1) of *UNCLOS* 1982, and Article 8(h) of the Convention on biological diversity 1992.

[77] The Convention will enter into force 12 months after the date on which not less than 30 states, the combined merchant fleets of which constitute not less than 35 % of the gross tonnage of the world's merchant shipping, have deposited their instrument of ratification with IMO as the depositary (Article 18). At 6.3.2013 there were 36 contracting states to the treaty representing some 29 % of the world's merchant tonnage.

[78] Different systems are based upon solid–liquid separation or disinfection, the latter employing either chemical treatment, such as chlorination, or physical processes such as UV radiation, de-oxygenation, ultrasonic treatment, cavitation etc. Lloyd's Register (2010), p. 7.

[79] Alternative schemes under development include the design of ships that rely on a process of continuous flow-through of ballast water or a Dutch design of a 'monomaran' hull, which does not require ballast water.

[80] Such alternative 'design out' rather than 'end-of-pipe' solutions would not only drastically reduce the threat of invasive species, but would also remove the need for investment in cleaning systems, solve the problem of corrosion in ballast tanks and reduce the turnaround time in port. Wijnolst (2010), p. 6.

3.2.7 *Abandoned Ships and the Nairobi Convention on the Removal of Wrecks 2007*

Two issues which may have a direct effect upon shipbreaking relate to the deliberate abandonment of ships by their owners, which may subsequently have to be dismantled, and the dismantling of ships that have been wrecked following an accident or collision or after grounding. In the former case, it has been determined that international legislation would not be the appropriate measure to address such instances; in the latter an international treaty was adopted at Nairobi in 2007.

After the removal of the crew, vessels may be abandoned by their owners for a variety of reasons, including accident and collision and lack of funds for maintenance and repairs. Some states also include confiscation and seizure of ships for illegal traffic of migrants or drugs, bankruptcy and other offences, and seizure by ports or other national authorities under the category of abandonment. The subject of abandonment was recognised as a matter of concern at the seventh meeting of the Conference of the Parties of the Basel Convention in 2004.[81] An outcome of a questionnaire produced by the Open Ended Working Group[82] was a note prepared by the Secretariat of the IMO, which determined that:

- The abandonment of a ship at sea for the purposes of disposal constitutes an uncontrolled dumping operation and a violation of the *London Convention and its 1966 Protocol*, and thus was subject to enforcement according to the procedures of the relevant Parties.
- The abandonment of a ship on land or in port, with or without its crew, was not covered by the *London Convention and Protocol*, but was a liability matter for the port state to pursue with the flag state and the owner, acts of abandonment in internal waters[83] of a state being a matter for that state to address according to its national laws.[84]

It was the decision of the first session of the ILO/IMO/BC Working Group on Ship Scrapping,[85] therefore, that the question of abandonment of ships on land or in port be a matter for the national laws of the State concerned, and *not* a matter for international treaty.

[81] Decision VII/27 COP-7, Geneva, 25–29 October 2004.

[82] As per decisions of the fourth session of the Open Ended Working Group OEWG IV/6 and OEWG IV/7.

[83] If a Party has chosen to apply the provisions of the London Protocol to its internal waters (under Article 7.2), then the dumping of a ship in that State's internal waters would be covered by the London Protocol.

[84] Basel Convention (2006).

[85] Held 15–17 February 2005 at IMO headquarters, London. International Maritime Organization (2005a).

The removal of wrecks did, however, become the subject of international treaty, under the *Nairobi International Convention on the Removal of Wrecks 2007*, but only with regard to wrecks deemed to be a hazard to navigation or which may be expected to result in harm to the marine environment or damage to the coastline or related interests of one or more states. The Convention defines a wreck as a sunken or stranded ship[86] *consequent to a maritime casualty*.[87] It may address some cases of abandoned ships at sea, when these ships are wrecks consequent to a maritime casualty and when they present an identifiable hazard.[88]

International provisions for the removal of abandoned and/or wrecked ships is distinctly limited and restricted to cases following the consequences of a maritime accident only, other circumstances being left to the provisions of the legislation of the relevant state(s). The removal of such vessels, whether abandoned or sunk or stranded, involves a measure of scrapping. However, many ships continue to be abandoned and it is considered that the coming into force of the *HKC* will not reduce this,[89] but may even promote further abandonments given the control over approved breaking facilities. The *Nairobi Convention* is not yet in force.

3.2.8 International Convention for the Safe and Environmentally Sound Recycling of Ships 2009 – The Hong Kong Convention (HKC)

The *HKC* emanated from a combination of factors, including a growing international concern with the practices of the major shipbreaking states; a failure to gain international agreement on the applicability of the provisions of the *Basel Convention* and *WSR* with regard to end of life ships; the lack of enforcement with regard to various international guidelines on shipbreaking; and the intense lobbying by NGOs, in particular Greenpeace and the members of the NGO Shipbreaking Platform.

The Convention, adopted in 2009, is not yet in force. Details of its development, its provisions and perceived lacunae are examined in Chap. 6.

---

[86] The Convention defines a ship as '*a seagoing vessel of any type whatsoever and include hydrofoil boats, air-cushion vehicles, submersibles, floating craft and floating platforms, except when such platforms are on location engaged in the exploration, exploitation or production of seabed mineral resources.*'

[87] The Convention definition of 'wreck' includes: a sunken or stranded ship; any part of a sunken or stranded ship, including any object that is or has been on board such a ship; any object that is lost at sea from a ship and that is stranded, sunken or adrift at sea; or a ship that is about, or may be reasonably expected, to sink or to strand, where effective measures to assist the ship or any property in danger are not already being taken.' Similar measures were incorporated into Indian maritime law, allowing stranded ships to be considered wrecks and thereby subject to demolition—see Chap. 5.

[88] *Basel Convention* (2010).

[89] Comment from discussions at IMO, 1.12.2009.

3.2.9 *Other International Legislation*

Other international agreements on the subject of the movement of toxic materials include the *Rotterdam Convention* 1998,[90] which came into force in 2004 and aims to ensure that exports of extremely dangerous chemicals only takes place with the agreement of the receiving country, and the *Stockholm Convention* 2001,[91] which aims to eliminate from commercial use 12 very dangerous global pollutants, including dioxins and furans. Both conventions regulate the trade in materials containing PCBs.

The use of asbestos has been banned in the USA since 1989, and in Europe under the EU Directive 87/217/EEC.[92] The use of asbestos in the construction of ships in all countries was banned from 2002 by revisions to the *International Convention on the Safety of Life at Sea* (*SOLAS*) 1974, Chapter II, and amendment 2000. Nevertheless, there are reports of asbestos still being found in new builds, with the material allegedly being used in Turkish and Chinese shipyards. The use of asbestos 'in thousands of gaskets and other seals' was found in the new chemical tanker *Caroline Essberger*, built by a Turkish yard and delivered to Dutch owners in 2009. The cost of removing this material from the ship was estimated as being 10 % of the original cost of the ship.[93] As well as direct introduction, asbestos may also find its way on board ships through the purchase of spares from many countries where asbestos is still used in manufacture.[94]

3.3 Regional Legislation

3.3.1 *European Waste Shipment Regulation (WSR) 2006*[95]

The provisions of the *Basel Convention*—including the *Basel Ban*—were transposed into European legislation via the 1993 *WSR*, which covers the control of shipments of waste within, into and out of, the EU.[96] As a Regulation, it is binding

[90] *The Rotterdam Convention on the prior informed consent for certain hazardous chemicals and pesticides in international trade 1998.*

[91] The *Stockholm Convention on persistent organic pollutants (POP)* 2001.

[92] *Directive on the prevention and reduction of environmental pollution by asbestos 87/217/EEC.*

[93] Osler (2010a), p. 1.

[94] Osler (2010b), p. 1.

[95] Initially *Council Regulation (EEC) 259/93 of 1 February 1993 on the supervision and control of shipments of waste within, into and out of the European Community – the Waste Shipment Regulation*, subsequently replaced by *Regulation (EC) 1013/2006 of the European Parliament and of the Council of 14 June 2006 on shipment of wastes*—applicable since 12 July 2007.

[96] This in turn was transposed into UK law by the *Transfrontier Shipment of Waste Regulations* 1994 SI 1994/1137 (as amended).

in its entirety and applies directly to all Member States without the need for further national enactment.

Council Directive 84/631/EEC sought to harmonize for the first time the control procedures for the shipment of hazardous wastes in the Community.[97] *Council Regulation (EEC) 259/93* (the *Waste Shipment Regulation*) transposed the international notification requirements for the movement of hazardous wastes and the minimization of such wastes as required by the *Basel Convention*, to which the Community has been a Party since 1994. The 1995 *Basel Ban* amendment was implemented by the EU (*Council Decision* 97/640/EC),[98] thus banning from 1998 the export of hazardous waste from the EU to non-OECD countries altogether.[99] In 2006, *Directive 2006/12/EC*, the *Waste Framework Directive*[100] established environmental protection safeguards during recovery or recycling of wastes, and planning, permitting and inspection requirements, together with the definition of 'waste'.[101]

The 1993 Regulation was subsequently replaced by *Regulation (EC) 1013/2006 on shipment of wastes* which simplified further the control procedures, incorporated recent changes in international law and strengthened the provisions on enforcement and cooperation between Member States in cases of illegal shipments.[102]

3.3.2 European Policy on Ship Dismantling

The European Union has taken a sustained, positive, and proactive stance on the promotion of ship breaking—usually referred to by the more neutral term of 'ship dismantling' within the European bodies. From the events relating to the loss of the tankers *Erika* and *Prestige* in 1999 and 2002 respectively and the subsequent and extensive pollution by oil of the Western European coastline that resulted from the sinking of those two vessels, Europe's approach to the phasing out of single-hulled

---

[97] The move to control the movement of hazardous wastes was initiated by the hitherto uncontrolled movement of toxic wastes from the Seveso incident in France in 1982, and also several cases where such wastes from Europe were being exported and dumped in developing countries.

[98] The ban is contained in Articles 34 and 36 of *Regulation (EC) 1013/2006*.

[99] End-of-life ships were not explicitly listed in Annex V of the Regulation, which defined the wastes falling under the ban. However, all but the most recently built of ships were likely to contain hazardous materials listed in Annex V, and therefore be covered by the export ban.

[100] *Directive 2006/12/EC of the European Parliament and of the Council of 5 April 2006 on waste*.

[101] *Waste Framework Directive* Art. 1a. Waste is 'any substance or object in the categories set out in Annex 1 which the holder discards or intends or is required to discharge.'

[102] Whilst green-list wastes may be shipped for recovery within the OECD like normal commercial goods and only accompanied by certain information, shipments of hazardous wastes and wastes destined for disposal are generally subject to notification procedures, with prior consent of all relevant authorities of dispatch, transit, and destination.

tankers, and the associated measures contained in the Erika I[103] and Erika II packages,[104] left the IMO somewhat at odds with the protracted pace at which the European legislative measures advanced.[105] Beginning with the American restriction of single-hulled tankers which resulted from the pollution caused by the grounding of the tanker *Exxon Valdez* in the (hitherto) pristine surroundings of Prince William Sound, Alaska in 1989, and the resultant restrictions contained within the US *Oil Pollution Act 1990*, the subsequent IMO and European measures to eliminate these vessels would require a major withdrawal of ships that would not fall within the new specifications for double-hulled tankers by 2010 and 2015.[106]

Europe's policy towards ship dismantling was first outlined within the 2006 *Green Paper on Maritime Policy*,[107] which supported the need for an initiative for international standards and clean recycling facilities, but recognised also that the EU might need, in addition, a regional set of initiatives to accommodate any gaps in the work being done at an international level and to bridge any gap until they became effective. The general objectives of the EU's strategy on ship dismantling are to ensure that:

> ...ships with a strong link to the EU in terms of flag or ownership are dismantled only in safe and environmentally sound facilities worldwide, in line with the [then] draft Ship Recycling Convention.[108]

In a speech to the European Parliament on 25 April 2006, European Environmental Commissioner Dimas reiterated Europe's classification of end-of-life ships as hazardous waste exports, prohibited under the Community's waste shipment regulations, and called for measures that would bring the international shipbreaking industry under just and protective measures of globally acceptable environment and

---

[103] The Erika I package was designed to enhance maritime safety through a series of measures including stricter port state control via a ban from all ports of the Union of ships older than 15 years that have been detained more than twice over the two preceding years, stricter monitoring of classification societies, and a speedier replacement of single-hulled tankers by double-hulled tankers. Also proposed for future action was the need for increased transparency on ship data, improved surveillance provisions and the formation of a European maritime safety structure. European Commission (2000a).

[104] Erika II called for further measures to bring about improvements in the protection of European waters from accidents at sea and marine pollution by more effective monitoring and control of sea traffic, the setting up of a compensation fund for oil pollution in European waters and the establishment of the European Maritime Safety Agency to provide technical assistance in training, monitoring and data gathering. European Commission (2000b).

[105] Horrocks (2003), p. 3.

[106] A peak of some 800 single-hulled tankers was expected to be phased out in 2010, although it was uncertain as to how many might be converted to double hulls or used for non-oil transport or storage purposes. See European Commission (2007, 2008). To some extent, the originally anticipated number of tankers withdrawn for scrapping was somewhat lessened due to transfers to trades in other parts of the world not under the direct control of European legislation. Dimas (2006).

[107] European Commission (2006).

[108] European Commission (2008), p. 5.

safety standards for the dismantling of ships, promoted by a responsible international shipping industry and by a European example of scrapping its own smaller ships—including naval and government-owned ships[109]—within its own Member States. The *Green Paper* also sought to re-establish some environmentally sound shipbreaking capacity within the Community.[110] The Commissioner paid tribute to the efforts contained within the UK's National Ship Recycling Strategy,[111] which at that time was already out for public consultation.

A full-scale industry able to demolish the bulk of European flagged vessels was not envisaged. Firstly, Europe does not contain sites of a capacity large enough to accommodate the largest of modern vessels, such as VLCCs and ULCCs, and secondly, and despite the protestations of those who oppose the application of *Basel* principles to end-of-life ships, Europe was not seeking the attract the industry away from the established sites of Asia,[112] recognising the importance of shipbreaking to those economies, but was seeking '*to ensure that minimum environmental and health and safety standards are observed worldwide*.'[113]

An earlier report undertaken on behalf of the Commission[114] by Det Norske Veritas (DNV) and Appledore International and issued in 2001 had concluded that, under current conditions, the move to develop a large and environmentally sound and economically viable shipbreaking industry in Europe from the marginal position it currently occupies[115] would be difficult to attain, due to the relative differences between Europe and South Asia in labour costs; health and safety provisions; the demand for the reusable and recyclable materials produced; and the consequently lower prices that could be offered to ship owners.

A Green Paper '*On Better Ship Dismantling*'[116] was issued by the Commission in May 2007, and put out for public consultation that month. The paper refers to European legislation on the export of hazardous waste, and a number of high profile cases of European ships going for recycling to Asia, but notes also the difficulties of actually implementing the appropriate legislation. In recognition of this difficulty, the Green Paper listed a number of options which might strengthen the task of

---

[109] Approximately 100 warships and other government vessels—mainly British and French—were expected to be decommissioned in the decade from 2007. Built between the 1960s and early 1980s, these ships contain relatively high quantities of asbestos and other hazardous materials.

[110] Dimas (2006).

[111] The strategy package, published on 27 February 2007, consisted of the UK ship recycling strategy, Overview of ship recycling in the UK—guidance, and UK ship recycling strategy—final regulatory impact assessment.

[112] '*The main purpose of this exercise is the protection of the environment and of human health; the objective is not to artificially bring back ship recycling business volumes to the EU, thus depriving countries in South Asia of a major source of revenue*.' European Commission (2007), p. 3.

[113] European Commission (2007), p. 4.

[114] Det Norske Vertitas and Appledore International (2001), p. 3.

[115] Current capacity in Europe and Turkey to deal with the demolition of European warships and state-owned ships was deemed to be adequate to cope with anticipated decommissioning of such vessels over the next decade.

[116] Dimas (2006).

enforcement. These include better controls by waste shipment and European port authorities, and targeting ships beyond a defined age or where other factors indicate that a ship might be ready for disposal. Enhanced port control would impact not only upon European-flagged vessels, but all vessels using the ports.[117] This in turn would require better co-operation and information exchange between the Member States and the Commission, using databases to identify such vessels and follow their disposal routes. This would also be enhanced by better co-operation with the recycling states and transit states, together with listed details of shipbreaking sites practising environmentally sound management.

Until the new international Convention, then being developed by the IMO, came to fruition *and met existing European standards* (author's emphasis), it was considered that no changes to European legislation would be made. Neither was there a need to change the provisions of the *Basel Convention*, or to grant exemptions to end-of-life ships, until an alternative international regime, offering at least equivalence to *Basel*, was in force.[118] It was also recognised however, that it was most unlikely that the new Convention would be available before the phasing-out of single-hulled tankers reached a peak in 2010. Whilst pushing for the earliest completion of the new measures,[119] other solutions would also be necessary to meet this demand. Such measures would include a strengthening of the EU shipbreaking capacity[120]; a transfer of technical assistance and technology to the recycling states; the organisation of a (voluntary) ship recycling fund by the shipping industry; and a European certification system for clean ship dismantling and an award system for '*exemplary green recycling*'.[121]

A resolution from the European Parliament in 2008 indicated (with a certain amount of finger pointing)[122] that, amongst a wide range of issues, it was considered:

[117] The 2007 *Green Paper* argues that the risk of enhanced control culminating in a major reflagging of ships from Europe may not be too high, since Europe remains an attractive venue for much of the world's shipping. A more likely scenario is one of non-European owners abandoning vessels in EU ports, to be dismantled at the cost of the taxpayers—see Sect. 5.1.

[118] In the public consultation exercise that followed the release of the Green Paper 2007, most of the stakeholders were in favour of global standards that were compatible with EU standards, whilst taking into account the needs of developing countries. European Commission (2007).

[119] Whilst respondents to the consultation document deemed it necessary to support the work of the IMO in producing a globally binding agreement, it was also recognized that support for EU membership within the IMO was very limited.

[120] Respondents to the consultation in general indicated the need to upgrade the existing facilities rather than increase European shipbreaking capacity.

[121] European Commission (2007), p. 15.

[122] European Parliament 2008 *European Parliament Resolution of 21 May 2008 on the Green Paper on Better Ship Dismantling (2007/2279(INI))*. The long preamble included such comments as '*whereas it is regrettable that possible action is discussed only after high-profile cases, such as the attempt by the French government to dispose of its aircraft carrier outside the EU…*' (preamble D) and '*whereas the Parliament previously…called for guidelines to be developed by the Commission to close this loophole* [*whereby ships deemed waste under Basel are also*

> ...ethically unacceptable to permit the humanly degrading and environmentally destructive conditions involved in the dismantling of ships to continue any longer,
>
> ...that the EU is partly responsible for the existing social and environmental problems
>
> ...ethically unacceptable that children are used by some dismantling contractors to do hard and hazardous work

and that the initiative of the analysis of the social and environmental problems arising '*is at least ten years overdue.*'

Parliament therefore called upon the Commission to tighten up, via guidelines and mechanisms, the prevention of evasion of the *WSR* and the classification as waste of ships due to be scrapped that do not meet the requirements of international conventions.[123] Amidst the many other calls upon the Commission, the call to negotiate with the IMO on the disqualification of beaching from the new convention was not successful.

The following year, the Commission issued a Communication,[124] intended to stimulate discussion and prepare the way for legislative proposals that will be initiated after the adoption of the new Convention that was (then) anticipated later in 2009. Action to urge the adoption of the new IMO convention was deemed by the EU to be an important influence on ratification and effectiveness of the convention, whilst a lack of action might be interpreted as implying a low priority on the issue. The Commission cites its positive involvement with *MARPOL* and the *Anti-Fouling Convention* as indicative of the way in which the EU has made its rules binding on Member States and has encouraged the ratification and implementation of international agreements by third countries.[125] Accepting, therefore, that the formal introduction of the new *HKC* might take an inordinate length of time, the Commission proposed that a start be made on:

> ...preparations for establishing measures on key elements of the envisaged Ship Recycling Convention as soon as adopted by the IMO diplomatic conference...in particular concerning surveys and certificates for ships, essential requirements for recycling facilities and rules on reporting and communication.[126]

Included within these measures is the possibility of shipowners using the services of breakers who can demonstrate improvements to their health and safety and environmental regimes. The cost of providing clean and safe facilities is estimated in the communication to be between US$50 and 150 per tonne of steel, which would be the reduced pricing available to owners. The Commission believes that '*European ship owners can be expected to act in a spirit of corporate social responsibility*,' and indeed some of them do, the example of the dealings of the A P Moller–Maersk shipping group with selected Chinese yards operating upgraded

regarded still to be ships under other international legislation] *during the revision of the Waste Shipment Regulation, but the Council refused to accept this...*' (preamble G).

[123] European Parliament (2008).

[124] European Commission (2008).

[125] European Parliament (2007).

[126] European Commission (2008), p. 6.

facilities is a well established partnership. However, it may well prove to be a step too far to ask owners voluntarily to accept a shortfall of up to half a million dollars for sending a 30,000 LDT ship to favoured breakers,[127] especially in the aftermath of a major recession in the industry.

In response, the European Parliament supported the Commission's Communication, urging that its views be included in the new Convention wherever possible, that ship recycling be regarded as an integral part of a ship's life cycle, and that a move rapidly be made beyond feasibility studies, with '*...concrete regulatory action at EU level that moves beyond the regrettably weak remedies of the IMO.*'

There is, therefore, some degree of frustration that exists between the EU and the IMO in their efforts to promote a safer hazardous waste and shipbreaking regime. The EU adopted an almost zealot's approach to reform shipbreaking (as in the case of single-hulled tankers), centred to a large extent on social responsibility as well as environmental protection. It has the advantage of being able to advance on a basis of regulatory authority over Member States in a manner that the IMO, with the necessary voluntary approach to a global clientele, is unable to employ, and although the efforts of the EU are, as it states, able to persuade others and move forward reforms, perhaps just a little more diplomacy in its diplomatic missives might yield more positive returns.

Meanwhile, public consultations,[128] communications[129] and conclusions[130] continued to appear from the Parliament and the Commission, with a focus more directly on the *HKC*, now that this had been formally adopted in 2009. With some 25 % of the world's merchant fleet flying the flags of EU Member States (despite the profusion of open registers) and about 40 % owned by European companies,[131] the EU sees itself as needing to take a major role in the new provisions and undertook an assessment of the links between the *Hong Kong* and *Basel Conventions* and the *WSR* in a manner similar to the Open Ended Working Group operating under the auspices of the IMO.[132]

---

[127] With what the Commission regards as a relatively small fall in sales price for a ship for disposal, plus the principles of producer responsibility and polluter pays, the question of public subsidies for developing shipbreaking capacity was ruled out. Instead, the Commission suggested that owners might be persuaded to used the more responsible yards by means of a new EU reward for '*exemplary ship recycling*' or a new '*Clean Marine Award*,' which would provide '*public recognition for recycling and shipping companies with a clear environmental profile*.' European Commission (2008), p. 7.

[128] European Commission (2008).

[129] European Commission (2010).

[130] Council of the European Union (2009).

[131] Council of the European Union (2009), p. 1.

[132] Whilst there might appear to be scope for a joint and concerted action between the two exercises, it is unlikely that the IMO would wish to extend its scope to include measures that are purely regional, rather than fully international.

However, at the sixth Annual Ship Recycling Conference held in London in June 2011, the Head of the European Commission's Waste Management Unit announced that it was doubtful as to whether the ban on exporting (from the EU) ships for scrapping outside the EU or OECD was enforceable since in 2009, only 9.4 % of ships under an EU flag were sent to OECD countries, the remainder[133] going to non-OECD breakers; the restriction was therefore not working.[134] The only exception to current rules applied to ships that had effectively been pre-cleaned prior to departure and this in itself usually rendered them unfit for navigation, leaving towing to the breakers' yards as an uneconomic option.

The EC welcomed the new *HKC* as an important step towards the safe recycling of ships and urged Member States towards ratification. The Convention was regarded as '*a cornerstone of EU strategy*' and the Commission regarded the maintenance of the existing regime as counterproductive in not setting up the right incentives, but a new replacement strategy was subject to the completion of an impact assessment then in progress.[135] The subject of a new European *Ship Recycling Regulation* (*SRR*) is discussed in Chap. 7.

3.4 National Legislation and Practices

There is a plethora of legislation, enacted by states, which encompasses the activities and the disposal of ships. Of particular relevance is that enacted by the legislatures of the major shipbreaking nations themselves; however, the extent to which these provisions (plus the international agreements to which the states may also be signatories) are actually and effectively regulated may vary considerably. This section will consider, as examples, just some of the legal instruments in force (but not necessarily enforced) in India and in Bangladesh with regard to shipbreaking. In both cases, the main contention appears to lie not between the breakers and the various authorities, but between those authorities (regional/local/national) and the respective High/Supreme Courts in enforcing the orders issued by the latter. It also illustrates the lack of overall policy as expressed by the extent to which, at times, various national ministries, especially in Bangladesh, appear to act in opposition to each other with regard to their shipbreaking industries. With a strong central control in China, and also in Turkey, the impetus there for reform comes essentially from central government, whereas in the Indian sub-continent the move for reform comes basically from the courts, which themselves have been heavily prompted by campaigning NGOs.

[133] If only large ships (as distinct from small coastal and fishing craft) were considered, the percentage going to non-OECD breakers was cited as being nearer 99 %.

[134] Burgues (2011).

[135] Burgues (2011).

3.4.1 India

India's *Hazardous Waste Rules 1989*,[136] were issued under t
the *Environment Protection Act 1986*, and were enacted to enfc
of the *Basel Convention*. In 1997, the import of all hazardous waste was pro...
by the Supreme Court, which created the High Power Committee (HPC) on Management of Hazardous Wastes[137] to monitor the enforcement of the rules. In its subsequent report on its investigation into wastes, the HPC included an assessment of the shipbreaking activities at Alang, and as well as commenting on the big contribution that the activities there made to the economy. It also commented critically on the '*very serious problems*' arising from the hazardous wastes and the health and safety problems arising from some of the activities involved,[138] noting that the environmental guidelines for the shipbreaking industry, issued by the Central Pollution Control Board, had never been implemented. The HPC subsequently found that the Ministry of Environment and Forests (MoEF) responsible for the enforcement monitoring had only four officers to oversee the implementation of the HW rules throughout the whole country, in addition to their other responsibilities. The State Pollution Control Board (SPCB) informed the HPC that they had not been allowed to recruit scientific and technical staff to meet their objectives.

The HPC consequently made a number of recommendations to address these issues. One of these resulted in the very specifically titled *Gujarat Maritime Board (Prevention of Fire and Accidents for the Safety and Welfare of Workers and Protection of Environment During Ship Breaking Activities) Regulations 2000*, which contain direct references to the norms and guidelines of various international conventions issued by bodies such as the IMO and ILO.

The Supreme Court also tasked the HPC with ensuring that a ship has all proper consents from the State Maritime Board or concerned authority stating that it has no hazardous waste or radioactive matter on board, and that all waste oil sludge, mineral oils, paint chips etc., are carefully removed from the ship and taken from the beach for disposal. No material should be burned on the beach and all hazardous substances that remain after decontamination should be removed and returned to the sender. Current practices do not indicate that these measures were followed, a view that was contained within its conclusion that the preservation of India's natural resources lay not with the statutory protection authorities, but in the hands of '*an alert and firm community*,' that was aware of the dangers arising and the adverse impacts they generated.[139] However, the report also stated that in the

[136] *The Hazardous Wastes (Management and Handling) Rules 1989*, amended 2000.

[137] Set up during a hearing of a Public Interest Litigation filed in 1995 by the Research Foundation for Science, Technology, and Natural Resource Policy, challenging the import of hazardous and toxic wastes into the country.

[138] Central Pollution Control Board (India) (2001a). Executive summary, p. 5.

[139] Central Pollution Control Board (India) (2001a). Executive summary, p. 10.

opinion of the GMB, '*no ship comes at Alang carrying with hazardous waste*' [*sic*] (*i.e.* as cargo).

The report also reminded OECD countries of their obligations to ensure that they should dispatch only ships that are '*hazardous/toxic contained free* (*sic*) *from every corner of the ship*' prior to being offered for sale. At the same time, it determined that since the Supreme Court order of May 1997 contained no reference to ship breaking it could not apply to the industry (a comment that many believe also applies to *Basel*). Further, it proffered that an empty ship coming to Alang cannot therefore be '*a Hazardous/toxic unit*,' thereby emphasising the need to pre-clean ships for scrap which might contain any traces of hazardous material whilst at the same time exempting them from the ruling.[140]

Central also to the debates and legal wrangling that have accompanied a growing number of high profile ships for demolition is the ruling of the Supreme Court of 2003 in response to the Writ Petition (Civil) 657 of 1995. In this ruling, a set of requirements relating to the demolition of ships was laid down and formed the foundation for subsequent cases of disputed scrappings—see Chap. 5. Of significance is the manner in which the requirements have been selectively bypassed or ignored when environmental and human health considerations have been pitted against commercial or economic needs, the latter usually managing to attain ascendency. The beaching and demolition of the former French aircraft carrier *Le Clémenceau* was ultimately thwarted only by its recall to France by the French *Conseil d'État* in the face of the potential political embarrassment to an imminent state visit by the President of France to India, despite much citing of the Supreme Court rulings. The case of the *Blue Lady*, a huge passenger liner that represented major employment and resource opportunities, was somewhat more successful (at least for the shipbreakers) in that permission to beach the vessel was finally granted by the Supreme Court, in the face of its own 2003 ruling, on what were cited as humanitarian rather than legal grounds. The Court's rejection of Intervention Application IA34 and the granting of permission to demolish the vessel, despite many references to the constitutional basis of Indian human and environmental rights, leaned heavily on the concept of the greater good, both to the local population and to the economy as a whole, in a way that might be applied to any subsequent challenge to restrictions on the import and demolition of end-of-life ships; this item is discussed further in Chap. 5.

There is, therefore, a certain inconsistency, and even contradiction, in the application or interpretation of court rulings within the regime to the extent that if the situation does not accord with the regulatory provisions, then the regulatory provisions may be amended to accommodate the situation. A prime example of this might be the redefinition of the term 'wreck' that arose in the case of the *Platinum II*—see Chap. 5.

[140] Central Pollution Control Board (India) (2001b).

3.4.2 Bangladesh

In Bangladesh, where practices and standards were generally deemed to be even lower than in Indian yards, there is a similar growing demand from both legal and political circles that ships should be decontaminated prior to export. For almost a decade, the situation in Bangladesh has been extremely volatile and yards have been ordered closed/open as arguments have taken place between breakers and NGOs, courts and government and also between rulings from the High Court and the Supreme Court. In 2003, the Bangladesh High Court ruled that the government of Bangladesh should ensure that ships for breaking should be imported in line with the provisions of the *Basel Convention* and there have been some successes, such as the refusal to accept the *Blue Lady* in the face of co-ordinated pressure from environmental NGOs—see Chap. 5. Such instances, however, have usually been the exception rather than the rule, as is true for the other major shipbreaking sites. The 2006 *Labour Law Act* contained provisions relating to working conditions, health and safety, hours, leave and compensation for the workers within the yards, but owing to a lack of political will and government resources, enforcement and compliance was considered to be almost non-existent.[141]

The following chronicle demonstrates the vacillations and tensions that prevailed around the Bangladesh shipbreaking industry and attempts to instigate observance of the state's legal provisions and court pronouncements. In September 2008, the Bangladesh Environmental Lawyers Association (BELA) filed a writ with the Bangladeshi High Court, challenging the entry to Bangladesh waters of the *MT Enterprise*, a vessel on the Greenpeace list of 50 'toxic ships' whose movements were to be reported.[142] In December of that year, the Department of Environment (DoE) had submitted a report stating that none of the 36 shipbreaking sites near Chittagong—all identified as category Red (extremely dangerous)—were operating with the required environmental clearance.[143]

In consequence, on 17 March 2009, the High Court directed the Department to ensure that all shipbreaking yards that were operating without environmental clearance were to cease operations within 2 weeks; this effectively shut down the industry. The Court also ordered that no ships should be imported for breaking without having been cleaned of hazardous materials before arrival. It further ordered that no ships on the Greenpeace list of 50 hazardous ships were to be allowed into the country, expressing its '*utter dismay*' at the way in which the ministries had failed to cooperate to ensure compliance with environmental laws, and the way in which the Department of Shipping had concentrated on importing more and more ships, at the expense of public interest, workers' welfare and environmental protection.[144] It further ordered the Ministry of Environment and

---

[141] Young Power in Social Action (2010).

[142] The Daily Star (2009).

[143] International Federation of Human Rights (2009).

[144] International Federation of Human Rights (2009).

Forest to compile within 3 months the necessary rules for shipbreaking, which observed the obligations of the *Basel Convention*, the domestic *Environment Conservation Act 1955* and the *Environment Conservation Rules 1997*.[145] A week later, following a petition by the Bangladesh Ship Breakers' Association (BSBA), the Supreme Court overruled the High Court order on closure, but other parts of the order were upheld.[146]

On 26 January 2010, the government issued a Statutory Regulatory Order (SRO) that required pre-cleaning certificates for all ships being imported for scrapping, confirming that a ship does not contain toxic materials.[147] The Court further ordered that any yards that could not certify that ships being broken were free of pollutants should be closed. The shipbreakers' response was to close their sites after the SRO was enforced.[148]

The following 2-year period was marked by constant changes and vacillations, including the formulation by the DoE of recommendations to impose some control over the industry[149]; a High Court ruling on the prohibition of importing ships without pre-cleaning and appropriate certification from exporting countries[150]; explosions bringing the industry to a standstill[151]; lack of permits allowing yards to resume operations and a national strike. The industry finally resumed operations in mid-January 2012.

---

[145] International Federation of Human Rights (2009).

[146] The Daily Star (2010).

[147] Khan (2010). In 2008, the country imported 200 ships for breaking and some 172 in 2009.

[148] The immediate effect of the strike was to halt the supply of scrap to the 250 local rolling mills—the shipbreaking yards are usually cited as supplying some 80 % of the nation's scrap. The breakers' counter-claim was that it was impossible to obtain pre-cleaning certificates for ships whose owners change, and that the move was the result of '*a global conspiracy to soil the prospective ship-breaking industry of Bangladesh*,' a conspiracy initiated by European countries also wishing to enter the industry. The strike was called off after 11 days in the hope that the SRO would be withdrawn, but instead, the restrictions were to be tightened. Khan (2010).

[149] These included restricting the industry to a specific zone to control its spread and its pollution; the requirement for ships to be pre-cleaned before arrival; details of hazardous materials remaining on board to be supplied to the MoE prior to ships being beached; liquid and solid wastes to be prevented from mixing with sea water during demolition. Alamgir (2010).

[150] The High Court ruling of 11 May 2010 further stated that No Objection Certificates (NOC) were not to be issued by the Department of Shipping without sight of such certificates. The Court also ordered the government to submit documentation on all the 172 ships that had been imported for scrapping since its order on 17 March the previous year, a ruling that was upheld by the Supreme Court in August. These legal prohibitions kept Bangladeshi shipbreakers out of the end-of-life ships market since the May ruling and seemed set to keep them out of the market for several more months. Such a removal of shipbreaking capacity had an impact upon the rest of the international shipbreaking industry, which was consequently able to lower prices offered for ships. It also obviously had a direct impact upon the Bangladeshi workforce employed at the yards, some 16,000–17,000 being without work as a result of the closures. McCarthy (2010).

[151] In January 2011, shipbreaking was again brought to a standstill after an explosion on the tanker *Pranam* reportedly killed four workers and possibly 26 others; a contempt of court order was issued against the yard owner for continuing breaking operations. Brown (2011), p. 9.

Whilst the debate on the reopening of the industry was evolving, the production of guidelines by the Industries Ministry was also undergoing a troubled gestation. The government had been directed by the Supreme Court to produce a set of rules by 14 December 2011[152] that were in accordance with a set of six environmental laws.[153] The *Ship Breaking and Ship Recycling Rules* 2011[154] that emerged were deemed to be biased towards the industry and inadequate to protect the safety of the workers or the environment[155]; the Rules were returned to the government for amendment and were re-presented on 8 January 2012 and forwarded by the Supreme Court to the High Court for evaluation. At the Ministry of Industries, a Ship Breaking Cell was established to monitor the implementation of the rules.

The 56-page *Rules* contain procedures for the inspection and approval of incoming vessels prior to permission being granted for beaching and demolition, and comprehensive provisions *for each site* for the identification, handling and storage of hazardous materials and training and certification. Welfare provisions for workers included the provision of clean drinking water, sanitation, changing and rest rooms, PPE, fire fighting and first aid facilities. Chapter VI contains specific provisions for compensation for injuries (including asbestosis) suffered by workers as the result of their work in the yards and the obligation of the yards to re-employ such workers in more appropriate functions, while Chapter III covers safety features such as the periodic testing of cranes and winches etc., the appointment of a qualified Safety Officer and emergency provisions. Special references are made to the handling of asbestos and ACM as well as ballast water and sludge and bilge water. No person under the age of 18 shall be employed and no female workers shall be allowed to enter the yards. The *Rules* also define procedures relating to the provisions and processes of the yard, including the preparation of ship breaking plans.

Control falls under the auspices of the Ship Building and Ship Recycling Board (SBSRB),[156] which is to work in co-operation with other government bodies with

[152] *The Ship Breaking and Ship Recycling Rules* 2011 were issued by the Ministry of Industries through a gazette on 12 December 2011. A second set of rules, *the Hazardous Waste and Ship Breaking Management Rules* 2011 was issued through a gazette by the Ministry of Environment and Forests on 22 December 2011.

[153] *The Territorial Water and Maritime Zone Act*, 1974; *the Basel Convention Act*, 1988; *the Marine and Fisheries Ordinance*, 1989; *the Environment Protection Act*, 1995; the *Environment Protection Rules* 1997; and the *Labour Act*, 2006.

[154] *Ship Breaking and Ship Recycling Rules* 2011. 'Formulated in pursuance of the Hon'ble (*sic*) High Court Division of Supreme Court, in writ petition No. 7260 of 2008 dated May 24, 2011 taking into consideration the directions contained in the order.'

[155] The Daily Star (2011). Contested provisions under the new draft rules included the requirement for importers to submit lists of hazardous materials that are 'on board,' omitting references to materials that are 'in built.' The import of warships, nuclear powered ships and large passenger ships was also to be allowed, in contravention of existing restrictions. Roy (2011).

[156] Defined in the *Rules* as a one stop service provider, under the Ministry of Industries. which also provides waste treatment and disposal plant, approves and supervises training, oversees all worker safety measures and evaluates and permits individual yards. *Ship Breaking and Ship Recycling Rules*, Chapter I, s. 3.

regard to inspections and which regulates the operation of the yards. Penalty provisions for offences are detailed in Chapter VIII, including a list of specified fines to be imposed upon '*wilful defaulters*' for a range of defined transgressions.

Although these *Rules* appear to echo the provisions of the *HKC* (as well as responding to many of the demands of NGOs) and should represent a major advance in the operation of Bangladeshi shipbreaking by addressing the manifold shortcomings of the industry, their success will be dependent ultimately upon the extent of political will and their observance and enforcement by both the breakers and the supervisory authorities if they are not to follow the fate of the *Labour Law Act*.

3.5 Shipbreaking Guidelines

A number of sets of guidelines have been compiled with the aim of controlling and improving the traditional processes of shipbreaking and ameliorating the impacts that they cause to human and environmental health. Three sets of guidelines, namely those compiled by the IMO, the ILO, and the Basel Convention, are in effect on an international, but purely non-mandatory, basis.

3.5.1 IMO Guidelines

In 1999, the shipping industry established a Working Party on Ship Recycling consisting of seven major industry organisations,[157] which in 2001 published its *Code of Practice* on ship recycling.[158] This was subsequently to become the basis for parts of the *IMO Guidelines on Ship Recycling*, which were finalised at the 49th session of the MEPC in July 2003, and adopted at the IMO 23rd Assembly in 2003.[159] The Guidelines were written to complement the ILO and Basel guidelines, as well as recognising the relevance of a number of other conventions such as the *Stockholm Convention on persistent organic pollutants* (*POPs*) and the 1972 *London Convention on dumping*.[160]

---

[157] BIMCO, ICS, INTERCARGO, INTERTANKO, ITOPF, ITF and OCIMF, all organisations with consultative status at the IMO.

[158] Marisec (2001). Subjects for particular mention included—'as far as reasonable and practical'—the safe removal and disposal of asbestos prior to arrival at the yard, the certification of tanks to a state safe for entry and hot work and the discharge of halon to approved facilities. The *Code of Practice* also called upon administrations of the shipbreaking states to issue Certificates of Approval to those sites with acceptable worker safety and environmental control standards.

[159] International Maritime Organization (2004).

[160] International Maritime Organization (2004), Application section 2.2.

The Guidelines were intended as guidance to all involved in the ship recycling process, and were therefore directed at a wide audience, ranging from shipowners, shipbuilders, flag, port and recycling states, shipbuilding authorities, intergovernmental organisations and suppliers of marine equipment etc., providing 'best practice' through the life cycle of a ship to the scrapping process. Conditions and practices existing in the scrapping facilities were to remain the obligation of the facilities themselves and of the regulatory authorities of the countries in which they operate,[161] although the Guidelines do address the need for the adequate control of wastes and the prevention of pollution in accordance with the *MARPOL* and *Basel Conventions*.[162] Whilst the Guidelines do contain a discussion of *Basel* and its provisions, they stop short of actually referring to the idea of an end-of-life ship itself being hazardous waste. The aims of the Guidelines were directed at the preparation of ships for recycling; the minimisation of the use of hazardous materials and the minimisation of wastes generated during a ship's operational life. They also sought to encourage co-operation between agencies and identified the problems of ship recycling as something that involved all stakeholders.[163]

Hazardous materials are to be identified according to three lists within the guidelines, the first being based upon Appendix B of the *Basel Technical Guidelines* and on Annexes 1 and 2 of the Industry code of practice. The Guidelines also saw the introduction of the concept of the 'Green Passport' for ships, an inventory of potentially hazardous substances built into the structure and equipment of the ship, which was to remain with the ship, details being updated as required. They also covered the preparation of a recycling plan, and the selection of the breaking facility in the light of its compliance with the *Basel* and ILO guidelines.[164]

The definition of a ship was given wide scope to include any type of vessel operating 'in the marine environment' including also fixed or floating platforms and vessels stripped of equipment or has been towed.[165] This definition bypasses any reference to an end-of-life ship being classified as hazardous waste and the Guidelines were categorised by BAN as a cynical ploy by the shipping industry to usurp some of the roles of the *Basel Convention*, but in a regime based on lower rather than highest common denominator. By pinning much of the responsibility for improvement on the breakers whilst by-passing the polluter-pays principle, BAN categorised the new Guidelines as:

---

[161] International Maritime Organization (2004), Introduction, section 1.7.

[162] International Maritime Organization (2004), Section 8 Preparations for ship recycling, 8.3.3–8.3.4 and Section 9 Role of Stakeholders and other bodies, 9.4.1–9.4.4.

[163] International Maritime Organization (2004), Introduction, section 1.6.

[164] International Maritime Organization (2004), Section 8 Preparations for ship recycling 8.1–8.3.2.

[165] International Maritime Organization (2004), Section 3, Definitions.

> . . .a precedent of allowing an industry to go forum shopping within the UN store for the weakest international law available [which] threatens not only the future of the Basel Convention, but the credibility of the entire UN system.[166]

These Guidelines did, nevertheless, represent an attempt to ameliorate conditions in the shipbreaking yards, principally by addressing those sectors of the shipping industry over which it had influence with regard to removing or limiting hazardous materials and hence the pollution that they caused as well as looking to the shipbreaking state to issue some form of official approval. At a subsequent session of the MEPC in 2005,[167] the Committee was asked to develop a new legally binding instrument to regulate the design, construction and operation of ships in a way that would facilitate environmentally sound ship recycling; this instrument eventually emerged as the *Hong Kong Convention* and incorporated many of the provisions of the *IMO Guidelines*.

3.5.2 ILO Guidelines

The International Labour Organisation has since 2000 had a focus on the health and safety issues of shipbreaking, although its wider concern with the industry dates back to the late 1980s. It had already issued a number of Conventions, Recommendations and Codes of Practice which, if not directly aimed at the shipbreaking industry, nevertheless were applicable, although little observed. In 2003, its Governing Body charged the ILO to 'draw up a compendium of best practice, adopted to local conditions, leading to the preparation of a comprehensive code on occupational safety and health in shipbreaking,' including the requirement for ships to carry and maintain throughout their lives inventories of hazardous materials on board.[168] The result was the '*Safety and Health in Shipbreaking: Guidelines for Asian countries and Turkey*'[169] which were aimed at both shipbreakers and competent authorities to provide guidance on improving the working conditions at the yards and to develop minimum criteria that would enable different facilities to be ranked. The Guidelines were intended to complement existing national legislation

---

[166] Basel Action Network (2006).

[167] MEPC 53. 18-22/07/2005 London, IMO. The Committee endorsed the views of the Intersessional Working Group that IMO should develop, as a matter of high priority, a new instrument to provide legally binding and globally applicable ship regulations for international shipping and for recycling facilities. Also agreed was a draft Assembly resolution setting out the IMO's commitment to develop a new instrument on ship recycling for submission to the 24th session of the Assembly. The objective is to complete the draft instrument in 2007 for its consideration and adoption in 2008–2009. IMO (2005b), p. 17.

[168] A conclusion of the Tripart Meeting on Social and Labour Impact of Globalization in the Manufacture of Transport Equipment, endorsed by the ILO Governing Body at the 279th session of the ILO, November 2000.

[169] International Labour Organisation (2003).

and other provisions, including the *IMO Guidelines*, the *Basel Convention* and *London Convention* and *Protocol* and the ICS *Industry Code of Practice*, whilst having no legal force themselves. Being dependent upon local circumstances, scale of operation and availability of finance, their non-observance might thus be effectively rationalised.

3.5.3 Basel Convention Guidelines

At the fifth meeting of the Conference of the Parties in December 1999, the Basel Convention's Technical Working Group was mandated to collaborate with the appropriate body of the IMO[170] to prepare guidelines for the environmentally sound management of the dismantling of ships. The resulting '*Technical Guidelines for the Environmentally Sound Management of the Full and Partial Dismantling of Ships*' were adopted at COP-6 held at Geneva on 13 December 2002 and issued in 2003. Primary movers were India, Canada and the Netherlands, whilst France and Turkey also became active as a consequence of their involvement in the *Sea Beirut* case—see Chap. 5. The Guidelines were based upon work by the Danish Environmental Protection Agency,[171] specifically undertaken to assist in the formulation of the *Basel Guidelines*.

The *Basel Guidelines* did not address measures to minimise hazardous materials aboard a ship prior to its despatch to the breakers, nor did they cover in any depth the ILO work on occupational health and safety, but concentrated on issues relating to the procedures and practices required at the shipbreaking facilities to attain Environmentally Sound Management (ESM), which it defined as:

> ...taking all practical steps to ensure that all hazardous wastes or other wastes are managed in a manner which will protect human health and the environment against the adverse effects which may result from such wastes.[172]

3.5.4 National Provisions

The American shipbreaking standards issued by the US Environmental Protection Agency[173] are regulated by an agency which can be stringent in its enforcement (despite past failures to prevent the export of ships for demolition elsewhere—see Sect. 5.8). However, the USA is a minor operator in the international shipbreaking

---

[170] Decision V/28. Representatives from the ILO, the ICS, and environmental NGOs also participated in the formulation of the Guidelines.

[171] Danish Environmental Protection Agency (2006).

[172] Basel *Convention* Article 2, para. 8.

[173] United States Environmental Protection Agency (2000).

scene with just a few operational sites whose activities are largely restricted to the disposal of government-owned naval and auxiliary ships.

In 2009, the Government of India's Ministry of Environment and Forests also issued a *Technical EIA Guidance Manual for Ship Breaking Yards*,[174] which contains details of the various national and international legislative controls and operational requirements that apply to the handling of hazardous materials, as well as details of the infrastructure requirements for the shipbreaking sites.

Various references are made to standards issued by the International Organisation for Standardisation (ISO), namely *ISO30000:2009 Ships and marine technology – Ship recycling management systems – Specifications for safe and environmentally sound ship recycling facilities* and *ISO1400:2004 Environmental Management Systems*, which are cited by a growing number of shipbreaking yards as evidence of their 'green' credentials, but these standards are commercial provisions and represent no legal requirement or standing.

References

Alamgir N (2010) Eight proposals to cut down risks. The Daily Star, 14 April 2010, Dhaka

Andersen AB (2001) Worker safety in the ship-breaking industries: an issues paper. ILO

Basel Action Network (2005) BAN comments and proposals for resolving Basel Convention shipbreaking issues. Submitted to the Joint ILO/IMO/BC Working Group on Ship Scrapping, 7 February 2005, IMO

Basel Action Network (2006) The shame of shipping; breaking the principle to break ships. Briefing paper, 5 April 2006. www.env.go.jp/g8/mats080524a/01ban-handout.pd. Accessed 20 Oct 2010

Basel Action Network and Greenpeace International (2002) Shipbreaking and the legal obligations under the Basel Convention. Submitted by the Basel Action Network and Greenpeace International for the fourth session of the Legal Working Group of the Basel Convention, 10 January 2002, Seattle

Basel Convention (2006) Implementation of the decisions adopted by the Conference of the Parties at its seventh meeting: ship dismantling. Conference of the Parties to the Basel Convention, eighth meeting, Nairobi, 27 November – 1 December 2006 Item 6 (f) of the provisional agenda. UNEP/CHW.8/7

Basel Convention (2010) Guidance on dealing with abandonment of ships on land or in port. Annex to the Report of the Open Ended Working Group, Further to Decision VIII/13. Open Ended Working Group, seventh meeting, Basel Convention, 10–14 May 2010, Geneva

Bhattacharjee S (2009) From Basel to Hong Kong: international environmental regulation of ship-recycling takes one step forward and two steps back. Trade Law Dev 1(2):193

Blankestijn TP (2012) Successful initiatives for implementing the standards of the Hong Kong Convention so far. In: 7th annual ship recycling conference, 29–30 June 2012, London

Brown H (2011) Yard blast halts all Bangladesh breaking. Lloyd's List, 21 January 2011, London

Burgues JC (2011) EU regulatory review. EU's ship recycling strategy. In: 6th annual ship recycling conference, 14–15 June 2011, London

Central Pollution Control Board (India) (2001a) Report of the High Powered Committee on management of hazardous wastes

[174] IL&FS Ecosmart Limited (2009).

Central Pollution Control Board (India) (2001b) Vol. II Annex AII Documents related to Alang shipbreaking activities
Council of the European Union (2009) Council Conclusion on an EU strategy for better ship dismantling 2968th Environment Council meeting, 21 October 2009, Luxembourg
COWI (2009) Support to the impact assessment of a new legislative proposal on ship dismantling (COWI A/S for the European Commission DG Environment). Final
Danish Environmental Protection Agency (2006) Implementation of green ship recycling. Pocket Book Manual. Draft B
Det Norske Vertitas, Appledore International (2001) Technological and economic feasibility study of ship scrapping in Europe. Report for the EC No. 2000–3527 Final
Dimas S (2006) Solutions for responsible recycling of ships. Speech by the European Environmental Commissioner to the European Parliament, Brussels, 25 April 2006
European Commission (2000a) Commission Communication of 21 March 2000 to Parliament and the Council on the safety of seaborne oil trade. COM(2000) 142 Final, Brussels
European Commission (2000b) Communication from the Commission to the Council and the European Parliament of 6 December 2000 on a second set of Community measures on maritime safety following the sinking of the oil tanker *Erika*. COM(2000) 802 Final, Brussels
European Commission (2006) Towards a future maritime policy for the Union: a European vision for the oceans and the Seas COM (2006) 575 Final EC, Brussels
European Commission (2007) Green Paper on better ship dismantling COM(2007) 269 Final, Brussels
European Commission (2008) An EU strategy for better ship dismantling. Communication from the Commission to the European Parliament, the Council, the European Economic and Social Committee and the Committee of the Regions. COM(2008) 767 Final, Brussels
European Commission (2010) An assessment of the link between the IMO *Hong Kong Convention for the safe and environmentally sound recycling of ships*, the *Basel Convention* and the EU *Waste Shipment Regulation* Communication from the European Commission to the Council COM(2010)88 Final, Brussels. http://ec.europa.eu/environment/waste/shipments/background.htm. Accessed 29 Aug 2010
European Maritime Safety Agency (2011) Workshop report. Workshop on ship recycling, 27 & 26 June 2011, Lisbon
European Parliament (2007) European Parliament resolution of 21 May 2008 on the Green Paper on better ship dismantling (2007/2279(INI)), Strasbourg
European Parliament (2008) European Parliament resolution of 21 May 2008 on the Green Paper on better ship Dismantling. (2007/2279(INI)), Strasbourg
Greenpeace (1999) Ships for scrap. Steel and toxic wastes for Asia. A fact-finding mission to the Indian shipbreaking yards in Alang and Bombay Greenpeace, Hamburg
Greenpeace (2003) Ships for scrap VI. Steel and toxic wastes for Asia. Findings of a Greenpeace visit to Darukhana shipbreaking yard in Mumbai, India, December 2002
Horrocks C (2003) Wedge driven between EU and the world. Lloyd's List, 13 November 2003, London
Hossain MM, Islam MM (2006) Ship breaking activities and its impact on the coastal zone of Chittagong, Bangladesh: towards sustainable management. YPSA
IL&FS Ecosmart Limited (2009) Technical EIA guidance manual for ship breaking yards. Prepared for Ministry of Environment and Forests, Government of India
International Federation of Human Rights (2009) High Court halts deadly shipbreaking on the beaches of Bangladesh. www.fidh.org/High-Court-Halts-Deadly-Shipbreaking-on-the. Accessed 15 Sept 2010
International Labour Organisation (2003) Safety and health in shipbreaking. Guidelines for Asian countries and Turkey. October 2003. Issued at the ILO's 289th session in March 2004
International Maritime Organization (2001) Recycling of ships: report of the Correspondence Group. MEPC 46/7, 18 January 2001, IMO, London

International Maritime Organization (2004) Guidelines on ship recycling. Assembly 23rd session. Resolution A.962 (23) adopted 5.12. 2003, Item 9.7 IMO, London
International Maritime Organization (2005a) Consideration of the work programmes of the pertinent bodies of ILO, IMO and the Conference to the parties to the Basel Convention on the issue of ship recycling. Abandonment of ships. Note by the IMO Secretariat ILO/IMO/BC WG 1/2/2
International Maritime Organization (2005b) Report of the Marine Environment Protection Committee on its fifty-third session. MEPC 53/24, 25 July 2005 IMO, London
International Maritime Organization (2006). www.imo.org/OurWork/Environment/PollutionPrevention/PortReceptionFacilities/Pages/Port-reception-facilities-database.aspx
Khan K (2010) Ship-breaking conundrum. The Daily Star, 10 March 2010, Dhaka
Lloyd's Register (2010) Ballast water treatment technology. Lloyd's Register, London
Marisec (2001) Industry code of practice on ship recycling. Marisec, London
McCarthy L (2010) Bangladeshi yards await outcome of inspections. Lloyd's List, 27 October 201, London
Osler D (2010a) New tanker found riddled with asbestos. Lloyd's List, 22 June 2010, London
Osler D (2010b) DNV surveyors detect asbestos in more new builds. Lloyd's List, 1 July 2010, London
Putherchenril TG (2010) From shipbreaking to sustainable ship recycling. evolution of a legal process. Koninklijke Brill NV, Leiden
Roy P (2011) SC directives ignored. The Daily Star, 12 December 2011, Dhaka
Secretariat of the Basel Convention (2003) Technical guidelines for environmentally sound management of the full and partial dismantling of ships
Secretariat of the Basel Convention (2004) Legal aspects of the full and partial dismantling of ships. compilation and summary of comments and observations to facilitate the work of the Inter-sessional Working Group compiled by the Secretariat. Basel Convention, 7 April 2004
Shrinivasan P (2001) The Basel Convention of 1989 – a developing country's perspective. Liberty Institute, India
The Daily Star (2009) Ship-breaking ordered shut. The Daily Star, 18 March 2009, Dhaka
The Daily Star (2010) Pre-cleaning report must before import. The Daily Star, 12 May 2010, Dhaka
The Daily Star (2011) SC orders amending ship-breaking rules. The Daily Star, 14 December 2011, Dhaka
Ulfstein G (1999) Legal aspects of scrapping of vessels. A study for the Norwegian Ministry of Environment. Dept. of Public and International Law, University of Oslo
United Nations Environment Programme (2002) Legal aspects of the full and partial dismantling of ships. In: Conference of the parties to the Basel Convention on the control of transboundary movements of hazardous wastes and their disposal. Comments by the Division for Ocean Affairs and the Law of the Sea, sixth meeting, 9–13 December 2002, Geneva
United States Environmental Protection Agency (2000) A guide for ship scrappers. Tips for regulatory compliance. EPA 315-B-00-001, Office of Enforcement and Compliance Assurance
Wijnolst N (2010) Time to go overboard with ballast problem. Lloyd's List, 8 September 2010, London
Young Power in Social Action (2010) Shipbreaking in Bangladesh. http://shipbreakingbd.info/story/46%20workers%20killed%20on%20the%20shipbreaking%20beaches.html. Accessed 15 Mar 2012

Chapter 4
Ship Registration, Owner Anonymity and Sub-standard Shipping

4.1 Introduction

Whilst perhaps initially seeming to be somewhat tangential to the subject of shipbreaking, this chapter seeks to present a link between the registration of certain end-of-life ships and the question of liability by examining flag state responsibility (or the lack of it), and the way in which certain flag states not only permit, but also actively encourage, the anonymity of shipowners, and hence the liability which they carry for their vessels. Whilst hiding the identity of ownership may be undertaken for valid (and quite legal) commercial reasons, the practice also facilitates those owners who do not wish to advertise their ownership of ships poorly maintained or crews poorly supported, or who have a history of ship detentions, as well as those with distinctly illegal or even terrorist intent. Many ships may arrive at the breakers with their former trading identity on view, however some will arrive with both names and IMO numbers painted out whilst others may undergo a whole series of identity changes during the disposal process.[1] The COWI report prepared for the European Commission recognised reflagging just prior to scrapping as essentially a means '*to evade certain legal regimes and responsibilities*.'[2]

This obfuscation is the product of both minimal regulatory enforcement by specific flag states/registers, and of actively promoting various financial mechanisms intended to obscure beneficial ownership, a path that many shipowners appear to take (especially those with single-ship operations) when finally despatching end-of-life ships to the breakers. The desire for anonymity indirectly acknowledges the liability that owners have—and seek to avoid—for their end-of-life ships. The entry of a ship onto a state's register—and thereby into its jurisdiction—also imposes a duty on that state to apply and enforce all the conditions and obligations of operation and ownership contained in the legal instruments which the state has enacted or ratified.

[1] See, for example the case of the former ferry *Ben Ansar* in Chap. 5.

[2] COWI (2009).

M. Galley, *Shipbreaking: Hazards and Liabilities*, DOI 10.1007/978-3-319-04699-0_4,

An unregistered ship, unable to fly the flag of any state, is a stateless entity with no legal rights on the high seas. An owner may register a vessel in almost any state of his choosing, depending on his objectives and motivation. The choice of options is wide and each register may offer its own legal and commercial benefits. From the outset however, it needs to be recognised that whilst a number of open registers may be worthy of the tag 'Flag of Convenience'—and a number of registers use the very term itself in their websites—this is by no means true of all open registers; there are good and bad open registers just as there are good and bad closed registers. However, the new *HKC* is based upon a comprehensive system of surveys, certificates, plans and approvals, which is to be operated between flag states and the shipbreaking states. The extent to which some flag states currently preferred by those sending ships for demolition will have the administrative ability, or even the will, to adopt and adhere to the new requirements will be an issue that is interesting to follow.

4.2 Closed, Open and Second Registers

Shipping registers are traditionally divided into the 'closed' registers of the more traditional maritime nations, and 'open' registers of states which, in the most general terms, tend to have a more relaxed regime of financial and regulatory controls and often with minimal requirements of owner nationality. What might be termed intermediate registers, the 'second' or 'international' registers, lie positioned somewhere in between the other two categories. The history of the growth of various registers is adequately documented over a range of sources, and only the briefest of outlines will be given here.

Ships may only operate under a single registry.[3] Registering a vessel involves entering the details into a state's public records[4] and is a crucial element of public international law. It confers nationality on a ship and establishes the jurisdiction of that state over the ship to prescribe rules of conduct and enforce sanctions on the shipowners.[5] It also entitles the ship to fly its flag state ensign to denote its registration and nationality (the term 'flag' often being used as shorthand for the flag state) Entry to a state's register in turn places a responsibility on the state to ensure that the condition of the vessel, its manning and its operation, are in

[3] In the case of bareboat charters, whereby the company hiring a ship has both technical and commercial management responsibilities for the ship, a vessel registered in one state is permitted to fly the flag of another state for the period of the charter. During this period, the operator can choose which register offers the greatest trading advantages, and the vessel is regulated by the charter registry; however, matters of title and ownership remain under the regulation of the underlying registry. Frendo (2000), pp. 384 and 385.

[4] According to international regulation, some ships may be excluded from a nation's register on account of their small size.

[5] Hosanee (2009), p. 17.

compliance with the state's own national laws and standards, as well as with any international treaties to which it is a party.

Other public law consequences may be to confer the right to diplomatic and naval protection. In addition to transfers of registration or ownership for normal commercial reasons, owners may change registration if ships of a specific state are considered unwelcome in the ports of other nations. Owners may also change flags to those of major naval powers in times of conflict in order to enjoy the protection that such states may offer, such as the transfer to American or British flags during the Gulf War in the 1990s.[6] The absence on the current anti-pirate patrols of supporting naval units from Panama, Liberia, Cambodia, St Kitts Nevis, Cyprus etc., is noted.

A ship that has no entitlement to fly a flag of registration has neither nationality nor the right to protection under international law,[7] the US Court of Appeals stating the situation somewhat more abruptly in *US v. Marino-Garcia 1982*[8] in that

> ...vessels without nationality are international pariahs. Flagless vessels...represent 'floating sanctuaries from authority' and constitute a potential threat to the order and stability of navigation on the high seas.

Recognition of the flag was basically codified under Articles 4 and 5 of the *1958 High Seas Convention* (HSC), which grants the right to all states to sail ships under their flags on the high seas, whilst giving each state the right to determine the conditions under which that may take place. This recognition was subsequently incorporated into the 1982 *United Nations Convention on the Law of the Sea* (*UNCLOS*), Articles 90 and 91.

Closed or national registers have often imposed conditions as to the ownership, manning, and even construction of a ship within that state.[9] Until 1919, the right to allow ships to fly their maritime flags was a privilege granted only to maritime states (*i.e.* those that were not completely land-locked); only after the First General Conference of Communication and Transit held in Barcelona in 1921 was this right extended to all States, with or without sea coasts.[10] Today, a vessel's nationality is

---

[6] Middleton (2007), p. 15.

[7] This principle was tested in *Naim-Molvan v. Attorney-General for Palestine* [1948] A.C.351, whereby an unregistered ship flying the flag of Turkey was arrested by a British warship on the high seas off the Palestinian coast in 1948 whilst trying to land Jewish settlers, and was forfeited by a Palestinian court. An appeal against the arrest and seizure of the ship was made on the grounds that the freedom of the high seas also applied to ships of no nationality, but was rejected by the Privy Council, who upheld the court's forfeiture of the vessel.

[8] *US v. Marino-Garcia* 1982 67 9 F 2d 1373 (11th Cir.1982).

[9] Prior to its amendment in 1988, the *1894 UK Merchant Shipping Act* defined a British ship purely by its ownership by British subjects or 'bodies corporate' subsequently being redefined simply as a question of registration. Ready (1998), p. 3.

[10] Middleton (2007). The rights of a land-locked state to sail ships flying its flag on the high seas was reaffirmed in Article 4.1 of the *1968 United Nations Convention on conditions for registration of ships*.

generally taken to relate to its inclusion in the register of a particular State, *R v. Bolden and Dean (1997)*[11] providing case law.

Up until the period immediately following World War II, the majority of the world's merchant fleet was registered with states operating closed registers; thereafter, there began a rapidly growing transfer of ships to the newly established open registers. Under an American initiative, Panama first established an open register in 1916 and increasingly became a preferred choice for American shipowners, who were subject to the restrictive, domestic regime imposed by the USA's maritime legislation, and who could benefit from the new register's more relaxed manning and taxation regime. During the war, the transfer of ships from the Allies' registers to Panama permitted war materials to be shipped under the flag of a non-belligerent. After the war, an upturn in trade and the entry of a large number of surplus war-time ships into the market both contributed to a growth in the number and sizes of open registers. In the 1950s, open registers accounted for some 4 % of the world's fleet and, by the mid-1980s, more than 30 %. By the beginning of 2009, over 73 % of the total merchant fleet was registered under foreign flags (mainly in open registers), with Panama accounting for 23.6 % of the world's tonnage,[12] and by 2011 some 86 % of new tonnage was being registered to open registers.[13]

The primary function of a number of open registers today appears to be to develop an income stream for the state rather than develop a well-regulated fleet; a secondary function may be to develop related maritime economic activities. Whilst requirements for registration differ between states, there are features that tend to be common amongst many of the open registers, including low taxation regimes; minimal ownership requirements; reduced manning requirements; an inadequate or absent maritime administration and a general absence of regulatory observance.

Notwithstanding this *laissez-faire* approach, some open registers may operate a system that is as professional as some of their closed register counterparts, if not more so,[14] yet the less reputable of open registers may continue to be attractive to those operators who intend to restrict their operations to selected and less regulated geographical areas. Open registers in general have been associated with the pejorative tag of 'Flags of Convenience' (FOCs),[15] and this continues to be the case

---

[11] *R v. Bolden and Dean* (*The Battlestar*) (*1997*) 2 Int. M.L. Plymouth Crown Court held that the yacht *Battlestar*, although owned by US citizens, with a certificate of American ownership and entitled to fly the American flag, was not, in fact, deemed to be registered in any country since its owners had not entered the vessel into the US public registration records.

[12] Institute of Shipping Economics and Logistics (2009).

[13] Bergeron (2011), p. 6.

[14] A number of major oil companies have registered their tanker fleets with open registers and maintained and operated those vessels to the highest of standard. Farthing and Brownrigg (1997), p. 191.

[15] The term Flag of Convenience is particularly associated with the International Transport Workers' Federation (ITF), which, since 1948, has been waging a campaign against these registers, maintaining a list of countries it considers to be FOCs—currently numbering some

from bodies or organisations such as the OECD and the European Union, who see open registers as a threat,[16] and use the term freely in their documentation, a resolution of the European Parliament recognising flag of convenience states as '*a major barrier to combating illegal exports of toxic wastes.*'[17] However, despite the liberal approach displayed by various open registers, there is a move amongst certain flag states to distance themselves from the label of Flags of Convenience by limiting the age of ships that they will accept into their registers.[18]

In a number of instances, the open register nations do not actually operate their registers as a national body, but are content to hand over the control of their maritime flag to a third party in exchange for payment, hence the Mongolian and Cambodian registers are operated from Singapore whilst International Ship and Aircraft Registries offers the services of the registers of Belize, Cambodia, Dominica, Vanuatu and Georgia from Cyprus, as well as the Cyprian register itself.[19]

A specific and contentious feature of open registers relates to the existence of a 'genuine link' between the flag state and the ship. Under domestic law, a direct link is the product of ownership, crew etc, whilst under international law the link relates to the effective control over the administrative, technical and social matters that the state enacts over the vessel,[20] as defined under the *HSC* 1958 and *UNCLOS* 1982.[21] However, since no sanctions for the absence of a genuine link have been defined, the matter again remains one for the individual states to interpret. The question of the need for a genuine link suffered a further setback in 1960 when the International Court of Justice (ICJ) gave an advisory opinion on the constitution of the Maritime Safety Committee of the Inter-governmental Maritime Consultative Organisation (subsequently the International Maritime Organisation),[22] ruling that the definition of the (minimum of) eight 'largest shipowning states' to be included in the

32—as well as a list of blacklisted companies closely associated with FOCs. The inclusion of states on this list is based upon the 'Rochdale Criteria' as defined by the Rochdale Commission of 1970. Rochdale Commission (1970).

[16] In an interesting twist, Ready uses the term Open register as a euphemism for Flag of Convenience. Ready (1998), p. 17.

[17] *European Parliament resolution on the EU strategy for better ship dismantling, Preliminary* C. 26 March 2009.

[18] Liberia and Vanuatu have an age limit of 20 years, Cyprus uses 17 years and the Bahamas 12 years, although older ships may be accepted after special inspection.

[19] It is somewhat noteworthy that the website address for this organisation is www.flagsofconvenience.com. International Ship's Register offers an even wider range of what it labels 'Ships Registry in Flags of Convenience' from its base in the Canary Islands—flags listed include Barbados, Belize, Canary Islands, Comoros Islands, Dominica, Honduras, Liberia, Marshall Islands, Panama, St. Vincent and Grenadines, Tuvalu and Vanuatu. International Ship's Register (n.d.)

[20] European Parliament (1996).

[21] *Geneva Convention of the High Seas* 1958, Article 5(1), *United Nations Convention on the Law of the Sea* 1982, Article 5.

[22] Advisory Opinion of 8th June 1960 on the Constitution of the Maritime Safety Committee of the Intergovernmental Consultative Organisation, International Court of Justice (ICJ Reports 1960).

14 members comprising the Committee should be based purely in terms of tonnage registered, an opinion that was distinctly unappealing to the European shipowners.[23] Consequently, Panama and Liberia took their place on the Committee and, as a result, international open registry flags became legitimized in international law.[24]

Positioned somewhere between closed and open registers, second or international registers have usually been established as a response by certain of the more traditional flags to reverse the exodus of ships to open registers and away from their own national registers. Such registers have enabled states to maintain some control over their fleets, with effective maritime administrations able to secure adherence to appropriate regulations and standards, whilst allowing them to operate in a lower cost environment, at the same time safeguarding also some of their nationally-based businesses, skills and employment.[25]

4.3 Promoting Anonymity

Although certain open registers appear openly to display their rebuttal of internationally accepted standards and legal instruments, it is not simply the operation of the registers themselves that can prove attractive to the more nefarious (as well as many upright) shipowners in masking the identity of beneficial ownership; it is also the manner in which they promote the various financial mechanisms that allow corporate identities to be obscured in a web of nominees and holding companies which might be spread across a range of jurisdictions. The OECD report of 2003[26] on ownership and control of ships concluded that for beneficial owners who wish to

---

[23] As a result of the advisory opinion, of the eight originally nominated Member states of USA, UK, Norway, Japan, Italy, the Netherlands, France and Germany, the latter two were replaced by Panama and Liberia.

[24] Stopford (2009), p. 672. Even the pronouncement of the ICJ did not precisely define the term 'genuine link,' and it was not until the advent of the UN *Convention on the Conditions for Registration of Ships 1986* (*Arts 7 to 10*), that the question was more fully addressed, with the Convention requiring an economic link between the flag state and those that own, manage and man a vessel; this Convention, however, is not yet in force. Other case law has yielded up differing interpretations, particularly in the case of fishing vessels being selectively registered to take advantage of fishing restrictions or quotas—a practice known as 'quota hopping.' The European Court of Justice took the view that registration alone was sufficient to establish nationality and a genuine link. Churchill and Hedley (2000), p. 27.

[25] The first of these registers to be established was the Norwegian International Ship Register (NIS, 1987), under whose flag ships were able to sail with third world crews. Norway was followed by others, including France (Kerguelen Islands); Denmark (DIS); Germany (GIS); Portugal (Madeira); the Marshall Islands; and the Isle of Man (UK). The effect of the rise of these second registers is seen to somewhat blur the distinction between open and closed registers. Couper (2005), p. 27.

[26] OECD (n.d.)

remain anonymous, setting up a web of corporate identities was both easy and cheap. Whilst some registers try to establish the true ownership of a vessel, others deliberately promote anonymity as a major advantage.[27] Such registers may be deemed worthy of the title of Flag of Convenience.

The principal way in which an individual's ownership may be hidden is through the use of bearer shares which, unlike normal registered shares, do not carry the name of the owner and may be transferred from person to person without money changing hands or details of the transfer being registered. They thus facilitate a high level of anonymity, especially when issued to private limited companies, and their use is positively promoted by a number of open registers. When private companies, which are based upon ordinary (as distinct from bearer) shares, are initially registered, some details of shareholders are usually required. At this juncture, beneficial owners can hide their identity through the use of Nominee Shareholders who act on their behalf. Since corporate bodies require at least one Director nominally responsible for the operation of the company, the use of Nominee Directors acting as legal intermediaries can also hide the identity of beneficial owners since some jurisdictions cannot (*i.e.* do not wish to) compel nominees to reveal the identity of true beneficial owners. Allowing corporations to act as Nominee Directors adds a further level of obfuscation.[28] In many instances, the use of trustees, trust companies, lawyers and company formation agents offers further anonymity for those wishing to establish private companies, especially in offshore locations, where their active role may be limited to the provision of a local address, providing nominee shareholders and directors, and acting as local agents, but all with limited or no real operational function, lawyers and notaries in particular claiming professional confidentiality.

In terms of corporate structure, anonymity is furthered by the use of private limited companies,[29] which may easily be converted to shell companies and which have no assets or activities, but whose role is solely to hold legal standing as a corporate body. Such organisations are usually available easily and cheaply off-the-shelf. The use of one-ship companies protects any other assets of the beneficial owner from claims against a specific ship. Holding companies may be formed to hold the shares of the single-ship company, whilst the operation of the ships is undertaken by one or more management companies, shares in these operations being in the form of bearer shares held by the beneficial owner. By establishing a complex corporate web of organisations and nominees, across a range of jurisdictions, whose main interest may be merely the collection of registration fees whilst offering a high level of anonymity, the identity of beneficial owners can be hidden to the extent that might defy most investigations—see, for example, Sect. 5.1.

[27] Alderton and Winchester (2002).

[28] OECD (2003), p. 9.

[29] For example—International Business Corporations (IBCs), Trusts, and less frequently, Foundations and Partnerships. IBCs exist to facilitate international exchanges, but are barred from doing business in the country of incorporation, and therefore rarely pay taxes or are required to produce annual reports.

Given that the USA is probably the State most focussed on the security of shipping arriving at its ports, there is a certain irony in the fact that the open register, which facilitates anonymity, was a creature of their creation.

As well as promoting their own ship registration services, many states also offer offshore company registration in parallel.[30] At the same time, many of the traditional jurisdictions will allow local subsidiaries of foreign corporations to register their vessels under their flags, thereby allowing ownership to be concealed. By registering with overseas dependencies which operate their own registers but also hold close links with the traditional home jurisdictions, owners may also enjoy the benefits of naval protection and diplomatic representation that the home state provides. In such instances, recognised international representation or support, coupled with hidden ownership, may prove to be a dual advantage. A strong association with a traditional flag may also serve to reduce the emphasis on the monitoring and port state inspection of vessels which might otherwise be amongst the more targeted groups.

Changing the identity of a ship is a legitimate and, by now, frequent and valid commercial procedure[31]; and often whole fleets might be renamed as takeovers and mergers take place, the basic concept being to promote the name of the companies involved. The quest for anonymity often comes to the fore when ships are finally sent for disposal. Often when a ship is purchased for scrapping, usually by cash buyers, she will undergo a change of name and registry under her new owners. There are instances, however, when this change of identity happens whilst the vessel is actually at sea and *en route* to the breakers. Whilst it is conceivable for a ship to undergo several re-registrations in order to obtain economic advantages, it is difficult to associate such a series of changes being undertaken for any purpose other than a simple hiding of identity (and hence of liability) once it had ended its operational life and was actually at sea and *en route* to the breakers.

Another option for the new owner of an end-of-life vessel is not to register the ship with any registry, *i.e.* the vessel remains essentially flagless. This would allow the provisions of the new *HKC* effectively to be circumvented, since flagless ships do not have to be sent to an authorised shipbreaking facility or to have the international ready-for-recycling certificate as required by the Convention.[32]

---

[30] Not untypical are the advantages offered by the Seychelles, which in addition to allowing up to 100 % foreign ownership for ship registration, also offers incorporation of IBCs, with a high level of anonymity and privacy; no taxes of any sort; no ties to the EU; no accounting or reporting requirements; and no public register of company officers. SFM (2011). Similarly, the tiny island of Kiribati in the Gilbert Islands (area 811 sq. miles. and a population of a little over 100,000) offers similar inducements, including no restriction on ownership of ships; not being mandatory to incorporate a company in Kiribati or to register through lawyers; trading profits and capital gains are not taxed; there is no restriction on the nationality of a ship's owners or crew. Kiribati Ship Registry (n.d.)

[31] The general cargo ship *Black Sea Trader* may hold the record for name changes, having had some 14 different identities by the time she was demolished in 2011. Robin des Bois (2012).

[32] Bockmann and McCarthy (2009).

A significant indicator of the use and attractiveness of specific open registers is the discrepancy that occurs between the prevalence of flags chosen for ships in active commercial operation and the flags chosen for ships destined for the breaker's yard. Using statistics of the top 25 flag states as a proportion of the world's tonnage in 2008, a report[33] compiled for the European Commission showed that almost 43 % of the operational gross tonnage traded under just four states; at the point of scrapping, Panama (22 %), Liberia (10 %), Bahamas (6 %) and Marshall Islands (5 %). Whilst 42 % of the ships arriving at the breakers were under the two largest registries, Panama (28 %) and Liberia (14 %), the remainder of the top 25 included states that did not appear at all in the operational list, such as Tuvalu (6 %), and Mongolia (3 %), with Comoros Islands and Cambodia 1 % each.[34]

The inclusion of Panama and Liberia as leading flags for ships en route to scrapping could be answered by the size of the large fleets that they operate commercially; the inclusion of the other named states appears to be more of a function of the low fees, low crewing standards, high anonymity and short term registration that these states offer as FOCs. In addition, the shipbreaking monitoring organisation Robin des Bois reports that some 30 % of ships arriving for demolition are controlled by a classification society that is not a part of the International Association of Classification Societies (IACS).[35]

4.4 Sub-standard Shipping

Ships belonging to large multi-ship fleets are often sold before they reach the end of their working lives and before they are ready for demolition, some ships, especially in the major passenger liner groups, being passed down to other group members catering for less demanding clientele, rather than being sent directly for demolition.[36] Ships actually sold for demolition come, in the main, from '*small ship-owners operating in niche markets*.'[37] One of the consequences of the way in which beneficial ownership and liability are so easily obscured and limited, especially through single ship companies, lies in the way that certain registers facilitate (if not directly encourage) the prevalence of sub-standard shipping. Sub-standard shipping is manifest not only in minimal levels of ship maintenance, but also in low consideration for crews, both in terms of numbers and of basic provisions and in regard to internationally defined standards, as well as poor shore-based management etc.[38] Writing in 2000, Őzçayir stated that although we had a global shipping

[33] COWI (2009), p. 20.

[34] Figures rounded.

[35] Robin des Bois (2011).

[36] Discussions with representative of cruise ship fleet, 30.6.2011.

[37] Response from Leyal Ship Recycling Ltd., Turkey to the public consultation on the EC Green Paper, 2007. European Commission (2009).

[38] SSY Consultancy and Research (2001), p. 6. See also, for example, Sect. 5.1.

industry based on private enterprise, there were few who were actually dedicated to the safety of ships, crews or the marine environment.[39]

A report commissioned by the OECD[40] recognised the growing concern at the relationship between shipping casualties and sub-standard shipping in the 1970s. When trading conditions were poor, expenditure on ship maintenance was an easy target for owners, yet when conditions were good keeping a ship in full employment also tended to relegate upkeep to a low level of priority. Despite the introduction of various international treaties relating to the safety of ship operations, their uptake and enforcement has not always been to the forefront of some states' priorities.

The turn of the century witnessed a series of maritime accidents that demonstrated numerous parallels. In 1999, the 24 year-old tanker *Erika* broke in two and sank off the Spanish coast in heavy weather. In February 2001, the 27 year-old tanker *Kristal*[41] also broke in two and sank off the Spanish Atlantic coast in heavy weather. Both of the ageing ships were registered in Malta and were in need of repairs to replace sections of corroded steelwork and both ships had been classed by the Registro Italiano Navale (RINA). It was, however, the breakup of the *Erika* and the resultant extensive oil pollution,[42] reinforced by the similar loss of the *Prestige* some 3 years later, that triggered a number of legislative measures[43] from the EU aimed at removing such ships from European waters, but put the required action mainly in the hands of shipowners, flag states, and classification societies. This left (at least for a while) other bodies such as charterers, cargo owners, and brokers free to pursue whatever cheap rates they could find from these old, sub-standard vessels which, in the case of tankers, often work at the dirty end of the product market. Even so, many parties were able to cover their risks by insurance unless wilful negligence or recklessness could be proven, owners also being protected from the

[39] Őzçayir (2000), p. 1. In 2000, Őzçayir was a maritime law consultant and member of the IMO Roster of Experts and Consultants.

[40] SSY Consultancy and Research (2001).

[41] Following the loss of the *Kristal*, a meeting of the IMO's Marine Environment Protection Committee (MEPC) called for a review of Resolution A.744 (18) on guidelines on the enhanced programme of inspections during surveys of bulk carriers and oil tankers in order to strengthen the effectiveness of the Enhanced Survey Programme requirements, with reference to surveys and repairs. IMO (2001).

[42] The *Kristal's* cargo of molasses was dispersed at sea. In his description of the loss of the *Kristal*, Langewiesche describes the cargo as '*carried on the cheap by ships that are typically one step removed from the grave*.' Langewiesche (2004), p. 9.

[43] The Erika I measures were completed just 3 months after the loss of the vessel and included substantial amendments to the *Port State Directive* to increase the extent of port state inspections within the EU; measures to ban the employment of single-hulled tankers within EU waters; and a tightening of the *Classification Society Directive*, including suspending recognition of classification societies deemed to pay insufficient attention to enforcing ship safety standards. The Erika II package established an improved system of Community monitoring, control and information system for maritime traffic, a fund for the compensation of oil pollution damage in European waters and the formation of a European Maritime Safety Agency.

penalties of extensive pollution occurrences by the provisions of the *Civil Liability Convention*.[44]

Ownership of the *Erika* was with a single-ship company registered in Malta, the beneficial ownership of which proved difficult to trace until the individual in question came forward to identify himself. The resultant court case in the Paris Tribunal[45] in 2008, arising from the loss of the *Erika*, overturned the *status quo* to the extent that parties which hitherto had enjoyed a general avoidance of penalties were now also deemed to be at fault, the charterer, Total SA, being included with the owner, managers and classification society RINA[46] as guilty parties. The group was collectively fined the sum of €192 million, with Total SA receiving an additional €375,000 fine for failing to take into account the state and the age of the ship before chartering it.[47] Malta, as Flag State of the *Erika*, escaped penalties. By comparison, the fines imposed on the owner and on the classification society were relatively limited.

The OECD's (2003) report[48] contained a case study based on the consequences of the *Erika* oil spill and its impact on tanker and sale and purchase markets and

---

[44] 1969 *International Convention on Civil Liability for Oil Pollution Damage*. SSY Consultancy and Research (2001), p. 5.

[45] A subsequent review by the French *Court d'Appel* in 2010 upheld the earlier judgement, which 'breaks new ground' in that it was one of the few cases where the sinking of a vessel was tried under criminal law and that the classification society was amongst the accused. The Court also concluded that while the owner may benefit from civil compensation under the Civil Liability Convention, it is not protected for its role under criminal law. Further, the operator is not covered by the terms of the CLC, neither is the classification society, which had an equally independent role. The owner, who had been made aware of the need for repairs to the heavy corrosion on the ship by the classification society but had obtained an Oil Pollution Compensation Fund certificate to be able to lease the vessel for further commercial operations. RINA, who expressed doubts about the state of the vessel, nevertheless allowed it to sail; RINA was deemed to have issued certificates of compliance without prior inspection on a number of occasions. Mink (2010).

[46] In 2003, Spain sued ABS for US$1 billion in the US District Court of New York for violations under Spanish law and a breach of duty under US law. In August 2012, New York's Second Circuit Court of Appeal found in favour of ABS, ruling that ABS owed no duty of tort against coastal nations to refrain from 'reckless conduct' in its role as classification society, adding that even if such a duty was owed, the conduct of ABS could not be defined as sufficiently reckless as to result in the damage subsequently caused. The reporting website, however, sees this judgement as one that still leaves the question of legal duty of classification societies to third parties as unresolved. Robinson and Cole (2012).

[47] Hollinger (2008), p. 7. Total SA also set up an Atlantic Coast Task Force with a budget of €200 million to clean up the polluted beaches. Total appealed the judgement, claiming that it could not be responsible since the *Erika* was sailing under the Maltese flag and, since the ship had foundered in international waters, the French courts had no jurisdiction. (Although the ship went down in international waters, it sank within France's Exclusive Economic Zone which extends 200 nautical miles from the French coast.) Total's appeal failed in the *Cour de Cassation* in September 2012. France 24 (2012) Such risks to the reputation of oil majors, strengthened by the big impact of the pollution caused by the *Deepwater Horizon* disaster in the Gulf of Mexico in 2010, has been reflected in the increasing preference for crude tankers of less than 15 years of age, whilst Total SA is now reluctant to charter tankers which are more than 10 years old. Matthews (2011), p. 3.

[48] OECD (2003).

commented on the wide differential in charter rates between modern ships (*i.e.* ships less than 15 years old), and older tonnage, that had resulted from the accident. Yet this widening of rates for vessels of differing quality may actually serve to promote the continued use of certain sub-standard ships precisely because of the attractiveness of their low rates to certain shippers. The report cites the European Commission's view that the continued use of '*ships of appalling condition*' were still used to transport oil was due to the lack of any incentive to woo charterers away from the use of low quality tonnage and that this lack of concern perpetuated the operation of such ships.'[49] Meanwhile, ships that have reached the end of their lives and are fit only for disposal can also present major problems when their owners seek to distance themselves from the liabilities of the hazardous materials contained within the structure of ships.[50]

One of the consequences of inadequate flag state control over the vessels over under their jurisdiction has been the development of Port State Control,[51] whereby port states carry out inspections against a ship's various certificates and the relevant flag state's own national requirements, thereby imposing standards upon ships which do not observe them voluntarily. This acceptance of Port State Control has been portrayed as the acceptance by a flag state of the direct intervention by another sovereign state into the matter of a ship being an extension of its home territory.[52] The growth of Port State Control has led to the formation of several groupings of port states into regional operational agreements—Memoranda of Understanding (MOUs)[53]—to co-ordinate the inspection of vessels being identified as high risk in

[49] SSY Consultancy and Research (2001), p. 10.

[50] SSY Consultancy and Research (2001), p. 10. The report contains a detailed listing of the potential costs that may accrue to all involved with the employment of such ships. Other owners of more compliant and better maintained vessels may collectively suffer a general lowering of charter rates through the operation of substandard ships as well as a general rise in insurance costs and port inspections. Managers, operators, classification societies, charterers, shippers and cargo owners may all be increasingly accountable for the consequences of substandard shipping, as the cases of the *Erika* and the *Prestige* have demonstrated, and suffer losses through loss of cargoes, late deliveries resulting from port inspections and detentions, higher insurance costs, etc. The marine environment may also bear the cost of pollution resulting from accidents to shipping.

[51] The establishment of Port State Control actually originates with the provisions of the *1929 Convention on the Safety of Life at Sea* (*SOLAS*), reaffirmed in subsequent revisions 1960 and 1974 and in the provisions of the *International Convention for the Prevention of Pollution from Ships* (*MARPOL*) and *the ILO Minimum Standards Convention*. Farthing and Brownrigg (1997), p. 192.

[52] Farthing and Brownrigg (1997), p. 193.

[53] The first of these arrangements, the Paris MOU, was formed in 1982 by the maritime authorities of 17 European States, Canada and the Russian Federation. Subsequently, some eight other similar organisations have been established along similar lines to cover the Black Sea, Caribbean, Gulf Region, Latin America, West and Central Africa, Indian Ocean, Mediterranean and Asia-Pacific regions. As part of the European response to the loss of the *Erika*, the process of targeting substandard ships by a points system was changed to one whereby such highlighted ships have to undergo mandatory detailed inspections, ships repeatedly failing such inspections ultimately being blacklisted from entering European ports. *Directive 2001/106/EC of the European Parliament and Council amending Council Directive 95/21/EC*.

terms of their seaworthiness. As a natural development of association between states, similar coordination between different MOUs has started to develop and as a consequence of the shared results of monitoring and special inspection programmes, the potential operating areas for substandard shipping have started to shrink. However, the role of port states and MOUs remains but a substitute for effective control by owners and flag states. The OECD 2001 report sums up the regime with the comment that a combination of an accommodating flag state and a lax classification society offers ample scope for the owners of substandard ships to contravene international conventions.

4.5 Attempts at Revision

The OECD's final report on transparency of ownership[54] considered that improvements to transparency could result from a combination of changes to both the corporate mechanisms that are directed at promoting anonymity and to improvement in the transparency of the shipping registers themselves. However, it acknowledges that it is this very anonymity that is the attraction to users and the revenue they generate for the providers of the registers in question. It recognises the many vested interests that are present and suggests that a way forward may be to promote confidentiality, as distinct from anonymity, which may be more acceptable to all parties than unilateral action by individual administrations. However, in the absence of any autogenesis of improvements from the administrations themselves, attempts have been made to regulate against some of the less reputable of the registries, both directly and indirectly.

In a move against open registers, a *United Nations Convention for Registration of Ships* was adopted in 1986,[55] intending to bring order to the question of registrations, and ultimately to phase them out, by highlighting the need for efficient and competent maritime administrations through a genuine link between owners and flag states.[56] This was the product of two studies undertaken by UNCTAD and authorised in 1979, on 'repercussions of phasing out open register fleets' and on 'legal mechanisms for regulating the operations of open registry fleets.'[57] The Convention requires that a flag state shall have an effective maritime

[54] OECD (2004).

[55] Geneva, 7 February 1986.

[56] 1986 *United Nations Convention for Registration of Ships*. Article 1. Objectives. '*For the purpose of ensuring, or…strengthening the genuine link between a State and ships flying its flag, and in order to exercise effectively its jurisdiction and control over such ships with regard to identification and accountability of shipowners and operators as well as with regard to administrative, technical, economic and social matters, a flag State shall apply the provisions contained in this Convention.*'

[57] Farthing and Brownrigg (1997), p. 194. The studies were the result of pressure on UNCTAD from developing countries, who believed that the presence of open registers provided a barrier to them developing their own shipping fleets and obtaining cargoes.

administration with appropriate arrangements for registration and enforcement; fix the conditions for granting nationality; ensure that the ships flying its flag complies with relevant international standards of manning, safety, pollution control etc. and ensure that those responsible for the management and operation of its ships are readily identifiable and accountable. The question of ownership proved to be a contentious issue, centred on the provisions deemed necessary for a flag state to exercise adequate control and jurisdiction, whilst the subject of manning nationality was a polarising issue between the national and open register states.[58] The Convention also requires flag states to ensure the accountability of owners or their representatives be established in the state and to have legal and financial accountability. Today, the Convention is still not in force, only 24 of the necessary 40 States controlling 25 % of the world's shipping having ratified or acceded to it.

A proposal was made in 1989,[59] and subsequently amended in 1991,[60] by the European Commission for the adoption of a Community fleet and a Community shipping register—EUROS—in an attempt to halt the move of ships from the registers of Member States to the open registers of non-Member States, and the spread of secondary national registers, which the Commission regarded as a distortion of competition.[61] Inducements included lighter and more flexible manning requirements, fewer Port State Control inspections in European ports, and state aid and possible financial support for the training of seafarers.[62] The register was to work in parallel, rather than instead of, the national registers, and on a voluntary basis, a ship flying the Community flag alongside its own national flag to indicate its inclusion in both registers,[63] but the idea was rejected by the Council of the European Communities. The proposal was therefore replaced by *Council Regulation (EEC) 613/91 on the transfer of ships from one register to another within the Community*, which was aimed at facilitating completion of the internal market and providing for the mutual recognition of certificates issued in accordance with international rules such as SOLAS and MARPOL.[64] The EU has sought to improve the overall state of shipping by stepping up the pressure for Port State Control inspections at its ports and by pressure on the open registers operated by states such as Cyprus and Malta, which were then wishing to join the Union. The formation of the Paris MOU on Port State Control was an early manifestation of this process.

---

[58] Farthing and Brownrigg (1997), p. 195.

[59] COM(89)266, OJ C263 16.10.1989, p. 11.

[60] COM(91)483, OJ C19 25.1.199, p. 10.

[61] Ready (1998), p. 33.

[62] Frendo (2000), p. 385.

[63] Compare with Article 4, General Provisions of the *1986 United Nations Convention for Registration of Ships* viz. 4.3—Ships shall sail under the flag of one State only 4.4—No ships shall be entered in the registers of ships of two or more States at a time, subject to the provisions of paragraphs 4 and 5 of article 11 and to article 12 (relating to bareboat charters).

[64] European Parliament (1996).

The international shipping industry fully operates as a global industry, whose fortunes have, at times, waxed and waned dramatically. 'Traditionalists' have fought hard against the spread of open registers—and appear to have lost. Open registers now account for more than half of the world's merchant fleet and, although many open registers may operate a maritime administration able to register and regulate their vessels as well as (and, in some instances, possibly better than) some closed registers, it is the more relaxed regulatory regime offered in relation to ownership, taxation and manning etc.(and hence cost savings), offered by others that has proved irresistible to shipowners, whilst any desire or even ability on the part of some flag states to recognise and effectively regulate international standards is in direct opposition to their very *raison d'être*. Further, the well-publicised advantages of financial mechanisms designed specifically to hide the identity, and thereby the liabilities, of ownership through offshore corporations, nominees, bearer bonds, shell companies and the like, have doubtless encouraged many owners, especially those owners of single-ship companies, in the practice of minimal consideration both for the state of ships and regard for the crew—hence the prevalence of substandard shipping. In periods of a low trade cycle, expenditure on maintenance is easily ignored; in times of a high trade cycle, it may also prove secondary (or very secondary) to the opportunity for profitable trading. Either way, many ships are run on a totally profit maximisation/upkeep minimisation basis. Attempts at formal revision to the existing systems via UN Convention or EU Directives on registration have proved to be still-born and it has taken serious instances of ship losses as demonstrated by the sinking of the *Erika* and *Prestige* to forward the improvement of substandard shipping. It is against ships deemed to be sub-standard in terms of a troubled history or a polluted end-of-life state that the campaigning NGOs have fought so determinedly.

In 2009, the *Hong Kong Convention* was adopted. The Convention is founded upon a comprehensive system of documentation, surveys and inspections, approvals and certification for a ship destined for demolition. Given that a number of open registers are unlikely to have the maritime administration necessary to support such requirements, it will be interesting to observe how many such registers become party to the Convention, how many will be able, or even wish, to observe the requirements of this regime, and how their current clientele will regard such registers once the Convention enters into force.

References

Alderton T, Winchester N (2002) Regulation, representation and the flag market. J Maritime Res 4:89–105

Bergeron S (2011) Ship registry performance is not purely academic. Lloyd's List, 6 July 2011, London

Bockmann MW, McCarthy L (2009) Flagless vessel loophole may dent recycling plan. Lloyd's List, 19 February 2009, London

Churchill RR, Hedley C (2000) The meaning of the 'Genuine Link' requirement in relation to the nationality of ships. A study prepared for the International Transport Workers' Federation, University of Wales, Cardiff
Couper AD (2005) Historical perspectives on seafarers and the law. In: Fitzpatrick D, Anderson M (eds) Seafarers' rights. Oxford University Press, Oxford
COWI (2009) Support to the impact assessment of a new legislative proposal on ship dismantling COWI A/S for the European Commission DG Environment, Final Report
European Commission (2009) Details of EC (and other) documentation on ship dismantling. http://ec.europa.eu/environment/waste/ships/index.htm. Accessed 20 Nov 2009
European Parliament (1996) The Common maritime policy, Chapter 2, The sea and navigation. Directorate-General for Research. Working Document Transport Series W14 EP, Strasbourg
Farthing B, Brownrigg M (1997) Farthing on international shipping, 3rd edn. LLP Ltd., London
France 24 (2012) Court upholds total conviction in 1999 'Erika' oil spill. www.france24.com/en/20120925france-appeals-court-upholds-total-conviction-1999-oil-spills-erika-environment-disasteratlantic. Accessed 10 Mar 2013
Frendo M (2000) The future of open registers in the European Union. Lloyd's Marit Commercial Law Q 3:383
Hollinger P (2008) France fines total £143 m coastal damage from oil spill. Financial Times, 17 January 2008, London
Hosanee NM (2009) A critical analysis of flag state duties as laid down under Article 94 of the 1982 United Nations Convention on the Law of the Sea. UN Division of Ocean Affairs and the Law of the Sea, Office of Legal Affairs
IMO (2001) Report to the Maritime Safety Committee. Sub-Committee on ship design and equipment, 44th session. DE44/19 26 March 2001
Institute of Shipping Economics and Logistics (2009) Shipping statistics and market review. 53(7)
International Ship's Register (n.d.) www.internationalshipsregister.org/2html. Accessed 18 Apr 2011
Kiribati Ship Registry (n.d.) Kiribati ship registry. www.kiribaship.com. Accessed 18 Apr 2011
Langewiesche W (2004) The outlaw sea. North Point Press, New York
Matthews S (2011) Oil majors reluctant to charter older tankers. Lloyd's List, 8 April 2011, London
Middleton J (2007) Admiralty education ship registration and the role of the flag. Federal Court of Australia 2007. www.fedcourt.gov.au/how/admiralty_papersandpublications20.rtf. Accessed 23 Mar 2011
Mink E (2010) Erika process; French Appeal Court pronounced judgement. Mainbrace No. 3 July 2010. www.blankrome.com/index.cfm?contenID=37&itemID=2271. Accessed 3 Oct 2010
OECD (n.d.) Regulatory issues in international maritime transport. Directorate for Science, Technology and Industry, Division of Transport OECD
OECD (2003) Ownership and control of ships. Maritime Transport Committee, Directorate for Science, Technology and Industry OECD, March 2003
OECD (2004) Maritime security – options to improve transparency in the ownership and control of ships. Directorate for Science, Technology and Industry Final, June 2004
Őzçayir ZO (2000) Flags of convenience and the need for international co-operation. Int Marit Law 7:111
Ready NP (1998) Ship registration, 3rd edn, LLP Reference Publishing, London
Robin des Bois (2011) Information and analysis bulletin on ship demolition #25. www.robindesbois.org/english/shipbreaking25.pdf. Accessed 12 Mar 2012
Robin des Bois (2012) Information and analysis bulletin on ship demolition #26. www.robindesbois.org/english/shipbreaking26.pdf. Accessed 20 Apr 2012
Robinson and Cole (2012) Second Circuit addresses the tort liability of classification societies to third parties in *Reno de España v. ABS*. www.rc.com/newsletters/Publications/2153.pdf. Accessed 10 Jan 2013

Rochdale Commission (1970) Board of Trade Committee of Enquiry into shipping. Cmnd. 4337 1970
SFM Corporate Services (2011) Seychelles offshore company. www.sfm-offshore.com. Accessed 18 Apr 2011
SSY Consultancy and Research (2001) The cost to users of substandard shipping. Prepared for the OECD Maritime Transport Committee
Stopford M (2009) Maritime economics, 3rd edn. Routledge, London

Chapter 5
Case Studies and Legal Judgements

Whilst the concept and provisions of the *Basel Convention* remained the subject of debate within the shipping industry generally, the provisions of the Convention had been recognised by a number of European national courts via the incorporation of those provisions into the *WSR*.[1] Consequently, there are a small but growing number of legal decisions in which prohibitions have been placed upon the movement of ships destined for the breakers, even if, as in the case of France, they appear to have been imposed somewhat reluctantly. The announcement in June 2011 that the European Commission recognised that the *Basel Convention* was effectively unenforceable against ships destined for the breakers does not, in itself, necessarily render it invalid, it merely renders it unenforceable.

This chapter will examine a number of cases of ships despatched for demolition that attracted high profiles through the extensive campaigning against their demolition without any prior decontamination; judgements against their departure by exporting (and some importing) states; the liabilities defined within those judgements and the subsequent actions that resulted. It will also examine the responses that were made by the receiving nations, in some cases favourable, in others quite the opposite. In some instances the period between the original attempt to send a ship for scrap and the commencement of dismantling extended over a number of years, demonstrating the extent to which those opposed to traditional scrapping practices are prepared to campaign for the improvements they deem essential, whilst the frequent changes of ownership, name and registration of ships between departure and arrival at intended demolition sites indicate the extent to which owners attempt to avoid liability for hazardous waste. During the numerous cases where ships destined for scrapping have undergone one or more changes of identity during their final voyages to the breakers, little or no thought seems to have been afforded by the receiving state as to verifying the one element of a ship's identity that remains unchanged throughout its life, namely its IMO number. This could

[1] *Council Regulation (EEC) 259/93 of 1 February 1993 on the supervision and control of shipments of wastes within, Into and out of the European Community.*

M. Galley, *Shipbreaking: Hazards and Liabilities*, DOI 10.1007/978-3-319-04699-0_5,

quickly and easily define the identity and nature of ships, for example the *Platinum II*, for which scrapping permission was sought.

The cases display common themes. From the owners, there are the attempts to avoid liability for the hazardous content of end-of-life ships; from the judicial point of view, most of the judgements of the national courts are based upon an examination of the relevance of the *Basel Convention*, since all the states involved are Parties to the Convention. Further, the European states and Turkey consider their own regional/national legislation—the *WSR* and Turkey's *Regulation on the Control of Hazardous Wastes*—since these are also based upon the provisions of *Basel*. Judgements of the ECJ were cited by the European courts in the formulation of their decisions. The other relevant factor is more one of exception. The cases illustrated were the result of intense lobbying by the NGOs to prevent the demolition of the ships in their arrival condition. These, however, are the exception, and their paucity only highlights the large number of other ships that arrive for demolition without challenge.

5.1 Sandrien 2000

The case of the *Sandrien* provided an early example of a court ruling that a ship for disposal that contained asbestos should be regarded as hazardous waste, in this instance a court in the Netherlands ruling that the ship must not depart for scrapping whilst in a contaminated condition. Such a decision, however, came at a cost to the state of origin, which ultimately had to bear both the extensive mooring and demolition costs, which the owner successfully managed to bypass.

The *Sandrien* was an aging chemical tanker,[2] built in 1974 and sailing under the registry of Bolivia. In August 2000, the ship arrived in Amsterdam, where inspection revealed that she was suffering from severe corrosion and that her watertight integrity was in question. During the inspection, papers on board indicated that the ship was to make a final voyage to India, where she would be scrapped. Because of the hazardous materials on board, and because there was no notification of the transboundary movements to the competent authority, the Ministry of Housing, Spatial Planning and the Environment determined that this would breach European legislation. Lawyers representing the owners claimed that the ship was to continue trading, that she was not headed for Alang, and that the asbestos onboard was an integral part of the vessel. It was decided by the authorities that the ship would be allowed to leave only after major repairs had been carried out. It was also determined that following the repairs, the ship's final voyage should be to a breaker's

[2] Lloyds List had reported that the *Sandrien* had spent a year detained in the Italian port of Augusta whilst trading as the *Maria S*, owned then by Panships and controlled by Giuseppe Savarese, the owner of the *Erika*, which sank off the French coast in 1999. Moored next to the *Sandrien* in Amsterdam was the problematic ship *Otapan* – see Basel Action Network (2002).

yard and permission to depart would be granted only under the conditions that she sail directly to the breakers, in fine weather and without cargo. The presence of asbestos, heavy metals and other toxic substances meant that the vessel would be regarded as hazardous waste and the owners would therefore be required to apply for an export licence prior to departure. The owners, however, proved difficult to find, having set up offices in a number of different countries and communicating only through lawyers.[3]

In February 2001, a further order was served upon the *Sandrien* which prevented her sailing since there had been no notification about the transboundary movement of waste to the competent authorities. The export of such a ship for scrapping would be a violation of the *WSR*, since it had not been properly emptied of the hazardous waste. There followed several court procedures, including an interim ruling by the highest administrative court in the Netherlands, the *Raad an State*, provisionally quashing the government's detention order. However, the government initiated a more in-depth court action to consider the matter, and on 19 June 2002 the Court of First Instance in the Council of State, The Hague, ruled that an end-of-life vessel not properly emptied of hazardous material should be regarded as hazardous waste[4] and '*an important step towards the recognition of corporate and state accountability in the management of ships for scrap*.'[5]

An appeal was brought before the Council of State by the appellant, Upperton Ltd. of Mauritius, contesting the decision[6] of the Minister of Housing, Spatial Planning and the Environment to prevent the *Sandrien* from sailing from The Netherlands to India (under the application of executive coercion) as being in contravention of both the *WSR* and *UNCLOS*. The Minister's decision was based upon the fact that Upperton was about to transfer the ship to India without having lodged notice as required by the *WSR*, claiming that such an action breached Article 26 of the Regulation and section 10.44c of the *Dutch Environmental Management Act* 1992. To this the appellant responded that the *WSR* was not applicable to the vessel and no notice of sailing was necessary since the ship should be categorised as a 'green list substance'[7] since it contained no cargo residues that could be deemed waste, hazardous or otherwise. The appellant's interpretation of the phrase '*other materials arising from the operation of the vessel*' was that it did not cover asbestos used in the construction of the vessel.

The court considered the provisions of the *WSR*,[8] Article 1 defining waste as:

[3] Greenpeace (2002a). The owner was subsequently defined as Upperton, a Mauritius-based company who operating behind a post office box number in Mauritius. Vlierodam Wire Ropes Ltd. (2004).

[4] *Council of State, The Hague, Upperton Ltd. v the Minister of Housing, Spatial Planning and the Environment. LJN number AE4310 Case number 200105168/2.*

[5] Greenpeace (2002b).

[6] Minister's decision of 23 March 2001. Upperton's objection to this decision were countered by the further decision of 7 September 2001.

[7] As per rubric GC030 of the green list.

[8] *Directive 75/442/EEC*, as amended by *Directive 91/156/EEC*.

> any substance . . .in the categories described in annex I which are disposed of by the holder, of which the holder intends to dispose of or of which the holder is required to dispose.

The judgements of the ECJ for Joined Cases *Arco Chemie Nederland and others* and *Palin Granit Oy*[9] were cited by the court in relation to the question as to:

> . . .whether a substance was waste must be assessed in the light of all the circumstances and taking into account the goal of the Directive. . .ensuring that the effectiveness of the Directive is not undermined.

The Minister's decision to halt the export had been taken on the basis that, in addition to any cargo residues, the ship contained a substantial quantity of asbestos. From interviews and documents it was clear that the ship was to be demolished in India and the Dutch Shipping Inspectorate had only given earlier permission to sail for this purpose. When ownership had passed to Upperton in September 2000, the ship was deemed to be in such poor condition as to preclude its operation as a tanker or freighter without extensive repairs and it appeared that the original owner had therefore disposed of the ship as waste. The fact that the new owners had entered into a contract with Hatimi Steels in Alang to scrap the ship was not disputed and the court therefore concluded that Upperton intended to dispose of the ship as waste.

Subsequently, the plaintiff submitted that he now no longer intended to scrap the ship (having let the contract with Hatimi Steels expire), but had contracted with Shiva Marketing to sail the vessel with a cargo to India for use there as a 'floating vessel,' necessary repairs for this voyage having already been started. This claim, unsupported by evidence that the plaintiff had undertaken any repairs, was dismissed by the Court, which upheld the Minister's decision that the ship was deemed to be waste.

The appeal that the ship fell under the provisions of rubric GC030 of the green list of wastes (Annex II of the *WSR*) was also considered to be unfounded. The heading of the green list states that:

> Regardless of whether or not wastes are included on this list, they may not be moved as green wastes substances if they are contaminated by other substances to an extent which a) increase the risks associated with the waste sufficiently to render it appropriate for inclusion on the amber or red lists, or b) prevents the recovery of the wastes in an environmentally sound manner.

Since the presence of asbestos on board was not disputed, the Council determined that this contamination qualified the ship for inclusion on the red list and could not be transported under the provisions of rubric GC030.

The next claim by the plaintiff was that the defendant was not authorised to apply preventative executive coercion to prevent the ship from sailing, since there was no danger of an offence likely to be committed in the near future and which

---

[9] *Arco Chemie Nederland and others* Joined cases C-418/97 and C-419/97 [2000] ECR I-7411, and in *Palin Granit Oy* case C-9/00 [2002] ECR I-3533.

could lead to serious damage.[10] However, the Council determined that '*the departure of the ship must be deemed to be a first step towards illicit traffic*' and the Minister's powers under Article 26 to prevent the ship sailing as a first step towards waste transfer were in accordance with power granted under the *Dutch Environmental Management Act* and the *General Administrative Law Act* 1994.[11]

The plaintiff then submitted that the Minister's decision was in breach of Article 211 of *UNCLOS*, which requires that Member States, who have drawn up special requirements to combat contamination of the marine environment by foreign vessels entering inland waters to load or offload, are to announce these requirements, informing the competent international organisation thereof. However, the defence argued that the Minister's decision to prevent the ship from sailing was based upon the potential breach of the need for prior notification of the shipment of waste so that all relevant authorities may take appropriate provisions to protect human and environmental health. The Council therefore determined that the requirements of the *WSR* with regard to prior notification did not constitute special requirements for the arrival of ships as per the provisions of *UNCLOS*. Similarly, the plaintiff's claim for the principle of equality also failed in that the Council determined that he had provided no evidence that the defendant had waived administrative enforcement measures in similar cases. The appeal was therefore dismissed by the Council as unfounded in a judgement dated 19 June 2002.

During the 18 months that the ship was impounded, some 20 Indian crew members remained on board, without pay and unable to return home.[12] At the end of 2001, the International Transport Workers' Federation (ITF) arranged for the safe repatriation of the crew, but the owners merely replaced them with a new crew, even though the ship was unable to sail in its current condition. The replacement crew was finally repatriated to India by the ITF and the Dutch government after this Council of State decision. Scrapping of the ship began in November 2004, with the ship being emptied of hazardous materials, fuel and 9,000 tonnes of cargo before being broken up in dry dock.[13] Because of the difficulty of tracing the owners, the €1.7 million cost was borne by the Netherlands Government and the city of Amsterdam and not by the owners.

---

[10] The appellant claimed that the procedure for taking action against illegal traffic was defined in Article 26 of the *WSR* and that different enforcement measures were therefore not possible.

[11] *Algemene wet bestuursrecht.*

[12] Even though the crew received no pay, it was reported that the crew had themselves paid the ship's dealers for the maritime education that they were to receive while on board against the promise from the owners—a group of ship dealers—who had promised them work and a future despite the fact that the ship was unseaworthy. Greenpeace (2002c).

[13] Demolition took place in the new Ecodock, the dock owners using the project to promote further work for a proposed €50 million environmentally friendly demolition yard at Eemshaven in the North of the Netherlands. Vlierodam Wire Ropes Ltd. (2004).

5.2 Sea Beirut 2002

As a complement to the *Sandrien*, the case of the *Sea Beirut* serves as a first example of an importing state—in this instance, Turkey—refusing to accept for scrapping a ship deemed to contain asbestos and other hazardous materials, and which should therefore be treated as hazardous waste.

After suffering an engine breakdown off the autonomous port of Dunkirk in December 1999, the Liberian flagged *Sea Beirut* was abandoned to the Dunkirk harbour authorities by her owners Lane Holdings SA, a 'postbox' company, through a formal letter of abandonment. The relevance of the *Sea Beirut* case is that it provided one of the earliest examples of a ship destined for the breakers actually being rejected by the intended state of import, following representation by the NGO Greenpeace with regard to the importer's obligations under the *Basel Convention.*

An examination of the 27 year old ship established the presence of asbestos. The abandonment, together with the on-board asbestos, effectively classed the vessel as hazardous waste under the *Basel Convention* and the *WSR* and required that the Prior Informed Consent procedure be followed if the vessel were to be demolished outside the EU. The ship's real owner was never traced, but the *WSR* requires that if the real owner is unknown, then '*the person having possession or legal control of the waste* (*the holder*) *becomes the notifier*'[14] and responsible for notifying all the competent authorities concerned.

Ultimately, the vessel was offered for sale by auction by the port authorities. The purchaser was the German company MSK, acting as a front for the Turkish shipbreaking company CEMSAN, and the *Sea Beirut* left Dunkirk under tow for Turkey on 16 April 2002, still flying the Liberian flag. Although the port authorities believed that the ship was destined for repair rather than demolition, the port lawyer had also acted for the purchaser CEMSAN, which was an organisation that operated shipbreaking rather than ship repair facilities in Aliağa. France thus became the state of export for the ship, but took no steps to follow the prescribed procedures.

After further investigation, Greenpeace advised all parties involved of the illegal status of the ship and boarded the vessel to take samples of the asbestos material. Similar samples were also taken by the Turkish authorities and both proved the presence of asbestos. As a result, the Turkish Environment Minister announced that Turkey would not accept the ship, which contravened the Turkish *Regulation on the Control of Hazardous Wastes* 1995, a law that was also based on the provisions of the *Basel Convention*, and since no prior notification of the status of the ship had been received by Turkey, the order was that the ship be returned to France for cleaning before re-exporting it for scrapping.[15]

An appeal to have the return decision of the Minister rejected was submitted to the Izmir second Administrative Court by CEMSAN Ship Dismantling Metal and

[14] *Waste Shipment Regulation*, Article 2 iii.

[15] Greenpeace (2002c).

Steel Industry Trade Limited Company in May 2002. Defendants were the Ministry of Environment and Forestry, Ankara, the Governorship of Izmir, and the Sub-Provincial Governorship of Aliağa. The summary of the Ministry's defence, as cited in the Court decision,[16] was that the import of all hazardous wastes is prohibited and asbestos is classified as hazardous waste under the Turkish *Regulation on the Control of Hazardous Wastes*; Customs procedures had not been completed for the import of the vessel; and permission for its dismantling had not been granted. The Regional Administrative court also rejected the appeal, adding that the ship had to be returned within 30 days.[17] The summary of the Governorship's defence was that asbestos on the vessel made it subject to the provisions of *Basel* and the *Regulation on the Control of Hazardous Wastes* and that no documents proving the removal of the hazardous waste from the vessel had been provided. The Sub-Provincial Governorship claimed that no inspection had been carried out on the vessel, but that all transactions (by the authorities) were legal and appropriate.

The arguments proffered by the plaintiff were numerous and in places appeared to be somewhat contradictory. CEMSAN claimed that:

- They were not notified of any analysis reports
- The ministry does not have the authority to send back wastes to the exporting country
- The ship does not contain waste material
- Export can be concluded following the removal of the two steam pipes containing asbestos
- Steam pipes of a similar type are contained in all vessels
- All wastes are removed from the ship during the importing procedure
- Due to the vessel containing 99 % metal it should be considered as metal scrap, the export of which is not prohibited
- The facts of the matter will be revealed following the expert analysis and, as a result of the analysis carried out, the company reached the conclusion that 'the vessel is not considered as hazardous waste and asbestos is not waste.'

In examining the case, the Court considered the provisions of the *Basel Convention* and the Turkish *Regulation*. Annex 1 of *Basel* lists asbestos under code Y36 as a substance defined as hazardous waste and subject to transboundary movement (control). Article 4 of the *Regulation* defines 'hazardous waste' as any substance under Annex 1 (of *Basel*, on which it was based) with Article 5a stipulating that '*the import of any kind of waste is prohibited, without prejudice to the provisions in Article 38*.'[18]

[16] Izmir 2nd Administrative Court, Case No. 2002/496, Decision No. 2003/1184.

[17] Greenpeace (2002c).

[18] Article 38 of the *Regulation on the Control of Hazardous Wastes* states that '*it is prohibited for wastes to be imported to areas under Turkish jurisdiction and to free areas. However, the Ministry can issue permits in accordance with published communiqués, for the import of wastes that are of*

Following the Greenpeace protest, three samples were taken from the vessel, which was then sealed (Customs procedures not having been carried out); two of the samples were analysed as containing high levels of asbestos, which is classified as hazardous waste. The decision to refuse entry to the vessel and order its return to France was therefore determined by the Court to be in accordance with the law. Claims by the plaintiff were that the differentiating criteria for waste and product are contained in Annex 1 of the Turkish *Regulation*, but the Court held that the asbestos was not a product; the ship was contaminated with asbestos and should therefore be considered hazardous waste.

In contravention of the *WSR*,[19] France refused to take back the ship since it remained under Liberian registry. The ship was last reported to be lying off Aliağa.

5.3 Kong Frederik IX/Frederik/Riky/Ricky 2005

The unauthorised departure of a former Danish ferry and its ultimate scrapping on the beach at Alang gives yet another example of the relative ease with which shipowners are able to circumvent the provisions of the *Basel Convention* to which the exporting and importing states are parties. It also illustrates with great clarity how the Indian authorities unilaterally interpret the provisions of that Convention in a way that is singularly advantageous to the shipbreaking interests and to the domestic asbestos market, despite government-to-government level requests from Denmark to return the ship to Denmark for decontamination and thereby observe the obligations of the Convention.

On the 16 March 2005, the 51 year old *Kong Frederik IX* left the Danish port of Korsør, allegedly to be put in service as a cargo ship in the Middle East. The ship had sailed without permission since it had been ordered by the Danish authorities to remain in Denmark until it had been decontaminated. Once at sea, the ship was re-registered and its name was changed to *Frederik*. At that point, the ship had been sold to a postbox company in St. Vincent and was managed by the Mumbai-based Jupiter Ship Management Company.[20] On April 23, the vessel finally arrived at Alang, by now under the name of *Riky* (sometimes referred to as *Ricky*) and registered, according to the Customs Division at Bhavnagar, under the flag of the Democratic People's Republic of Korea (North Korea), but according to another source, under the flag of the fictitious country of Roxa.[21]

Whilst the vessel was at sea, the Danish Environment Minister Hedegaard had written to her counterpart, the Indian Environment Minister Raja, warning him of

economic value and have sectoral importance, provided that it is documented that such wastes will be used for health, research purposes'.

[19] Article 26 of Regulation 259/93.

[20] Dutta (2005).

[21] India Together (2006).

the imminent arrival in India of this ship-for-scrap vessel and the hazardous materials on board.[22] She advised him that the ship was carrying asbestos which, according to both the *Basel Convention* (Art. 2 para 1) and Danish legislation, must be characterised as waste, since the owners intended to dispose of it. Further, Article 9 of the Convention made transboundary movements of hazardous substances without prior notification an illegal traffic; Denmark, as the state of export, therefore had an obligation to ensure that the waste be properly taken care of, either by re-import to Denmark or by environmentally sound disposal elsewhere. She further reminded him that the Indian Supreme Court had set out rulings[23] to avoid the dumping of hazardous substances in India, which should '*participate with a clear mandate for the decontamination of ships of their hazardous substances such as asbestos etc. prior to export to India for breaking.*' She therefore requested his co-operation in denying permission to dismantle the ship and in returning it to Denmark for de-contamination.

In his response, the Indian minister stated that India, as a party to the *Basel Convention*, had actually '*...strengthened its national legislation Hazardous Wastes management notified in 1989 to ensure compliance of our obligations under the Convention.*' Strict compliance with the Supreme Court's directions was claimed. However, it had been determined (by India) that the ship could not be classed as waste within the scope of Article 2.1 of the *Basel Convention*, since:

> According to the Gujarat Maritime Board, the Gujarat Pollution Control Board, and the Central Pollution Control Board who have inspected the vessel there was no objectionable hazardous material on board ... only in built hazardous material which are part of the structure of all ships.[24]

Permission for beaching was therefore granted. The minister also added his assurances that '*India has adequate capacity to ensure environmentally sound disposal of the said ship.*'

The Supreme Court had noted that, before a ship arrives at a port, it should have proper consent from the concerned authority; the Chairman of the Supreme Court Monitoring Committee announced in unambiguous manner to the Chairperson of the Gujarat Pollution Control Board that the '*Riky must be mercilessly driven out of Indian sovereign territory without any further loss of time.*'[25] The Supreme Court had also ruled that India should participate in the Convention's Technical Working Group with a mandate for decontaminating ships prior to their export to India. By allowing the importation of the vessel without requiring details of its contents from the owners, despite the warnings from Denmark or issuing authorisations to either

[22] A copy of the letter from the Danish Environment Minister Connie Hedegaard to Mr A. Raja, Minister for Environment and Forests, dated 15 April 2006, is available via a link from the Greenpeace web site Greenpeace (2005a).

[23] *Writ petition no. 657 of 1995 Research Foundation for Science Technology National Resource Policy versus Union of India & Anr. SLP C No. 16175/1997 & CA No. 7660/1997.*

[24] Basel Action Network (2005).

[25] Greenpeace (2005a).

the importer or the owner, it was considered that the Ministry was showing a disregard for both the *Basel Convention* and its own legal system.[26]

The Indian Minister's assurance of the ability to ensure sound environmental *disposal* (as distinct from recovery) of the ship may be interpreted as a recognition of the ship as waste under the Convention definition and thereby puts it in contradiction to his previous assertion that the ship cannot be so classified under the Convention.[27] Further, since the fact that asbestos was not disputed to be a hazardous material, then its presence on board the ship renders the ship hazardous waste under the *Basel Convention*. The Basel Decision II/26,[28] sought to remove any ambiguities as to a ship becoming waste by noting that '*a ship may become waste as defined in article 2...and that at the same time it may be defined as a ship under other international rules.*'

Decision VII/26[29] also recognized that '*many ships...are known to contain hazardous materials and that such...may become hazardous wastes as listed in Annexes to the Basel Convention*,' reasoning that even if India chose to defy this decision, their failure to treat the ship as a *Basel* waste was incorrect. The Convention defines hazardous waste in two ways—firstly, by the provision of the Annexes to the Convention and, secondly, by the fact that any Party involved in a transboundary movement considers the shipment to be a hazardous waste by their own national law. Consequently:

> ...the fact that Denmark considers the Ricky to be a waste and a hazardous waste was a sufficient legal basis for India to be obliged to consider it as such as well.[30]

This provision within the Convention was specifically designed to prevent any state from unilaterally declaring waste to be not waste, thereby leaving other states exposed and vulnerable. Since the ship left Denmark without proper notification, this renders it also illegal traffic within the terms of the Convention.[31]

A second letter from Ms. Hedegaard was similarly ignored, and a letter from the Danish Foreign Minister to his Indian counterpart also failed to produce the requested action. The classification of the asbestos (and other materials) by simply one of the Parties involved in the shipment was sufficient to render it hazardous waste under *Basel*; under the provisions of the Convention, Denmark therefore had the opportunity of a settlement by reference to the Implementation and Compliance Committee, but its lack of action presumably indicated that it regarded such an action as being not really worth the cost or effort in the face of what had become an established practice in India.

---

[26] Dutta (2005).

[27] Basel Action Network (2005).

[28] Seventh Conference of the Parties, October 2004.

[29] Seventh Conference of the Parties, October 2004.

[30] As per *Basel Convention*, article 1.1.b. See also Basel Action Network (2005).

[31] *Basel Convention*, Article 9.

A joint thesis issued in 2006, however, puts the positioning of the parties involved in different contexts, using the technique of Narrative Policy Analysis and presenting the situation as one that can fall between different sets of international regulations and thereby '*can be characterised as a "floating" regulative situation open for interpretation.*'[32] By examining the actions of the various actors in the light of different ideologies—Ecological Modernization or Promethean Discourse or Sustainable Development—alternative scenarios are possible, the focus of the Danish Environment Ministry on compliance with existing regulation, for example, being re-presented as a story of the Minister's future work for new international regulation—hence the later assertion by the Minister that taking further action against India could damage the process (of the *HKC*) that was already under way at the IMO[33] Such an in-depth examination of events can be a useful tool in situation analysis, but is somewhat beyond the scope of this present research. At the same time, the onus for controlling exports subject to Basel lies principally with the exporting state; in attempting to re-import the vessel, the Danish Ministry was demonstrating appropriate due diligence.

Dismantling of the vessel was subsequently allowed by the Supreme Court Monitoring Committee on Hazardous Waste (SCMC), subject to the presence of officers from the CPCB, and GPCB. The joint inspection team had visited the ship 3 days after its arrival; an inventory of cargo materials prepared by the Department of Customs had noted that the ship owners could not provide a detailed inventory of in-built materials of the ship. It also noted that the workers were little equipped to handle asbestos waste expected from the ship and were not even properly trained. The team recommended that a complete inventory of hazardous materials built in the ship be required from the owner for the subsequent verification of disposal of the wastes and that this be made a compulsory requirement for the owners to submit to the GPCB well in advance to allow the necessary decision to be made regarding beaching.[34] By conducting its inspection *after* beaching, the GPCB had violated CPCB's guidelines requiring an inspection and inventory *prior to* beaching, this sequence being the very reasoning behind making a measured judgement on the granting of beaching permission. The inspection team's conclusion was that when the determination of the presence of hazardous waste required physical verification:

> ...the present system of checking by GPCB, after beaching practically hardly serves any purpose'.[35]

An Intervention Application (IA) filed with the Supreme Court on 13 August 2005 by the NGO Corporate Accountability Desk[36] advised of the violations made

---

[32] Krogstrup and Arleth (2006), p. ii.

[33] Krogstrup and Arleth (2006), p. 62.

[34] India Together (2006).

[35] India Together (2006).

[36] The Corporate Accountability Desk is an Indian NGO that seeks to hold polluting and rights abusing organisations accountable for their actions.

by the Ministry involved in allowing the *Riky* to beach, including violation of the Court's own orders of October 2003. A refuting affidavit was filed on behalf of the SCMC by an Additional Director at the Ministry who was also a member of the SCMC, demonstrating a conflict of interest between the MoEF and the SCMC, which was supposedly an independent committee created by the Supreme Court to oversee hazardous waste regulation in the country and to assist the Court in such matters.[37] The committee appeared to have become a mere subjunct to the MoEF, but this is the result of a system whereby the apex court usually asks the government to constitute its committees, which are then filled with members of the relevant Ministries, thus negating the basic principle of impartiality. The SCMC was subsequently provided with its own legal counsel. Scrapping of the *Riky* was completed in November 2005.

This catalogue of failures to observe orders from the Supreme Court with regard to the inspection of ships, their contents and their documentation, together with the unilateral interpretation of the provisions of *Basel* and the negative response to calls from foreign governments anxious to help uphold the provisions of an international convention to which both countries were Parties, does not bode well for the well-being of those directly involved in the actual breaking, their environment, or the new ship recycling convention. The failures also demonstrate the conflicts of interest between official bodies and the lack of social justice, coupled perhaps with a shipbreaking industry able to demonstrate its political strength in maintaining business-as-usual. The inherent weaknesses of the system were to be effectively demonstrated again in the case of the *Blue Lady*, considered below.

On the 4 May 2005, the Danish Environment Minister again wrote to Raja that two more Danish ships, the *Droning Margrethe* and the *Rugen* had left for India, but had 'disappeared' *en route*. Both ships constituted cases of illegal exports of hazardous wastes. The *Droning Margrethe* later reappeared as the *Beauport II*, and was allowed into India for demolition at Alang.

5.4 Le Clémenceau/Q790 2003–2009

The case of the *Clémenceau* was a clear illustration of the way in which a number of European countries, as well as Turkey, have made a stand against the import (for scrapping or pre-cleaning) of a ship reportedly carrying high levels of toxic substances. It illustrates also the conflict that existed between the various legislative and regulatory bodies of the penultimate importer, India. The final prohibition by the French *Conseil d'État* against the ship's demolition in India may have been a recognition of the relevance of the provisions of the *Basel Convention*, but may

[37] In 2002, the MoEF was fined Rs 10,000 by the Supreme Court for failure to comply with its orders regarding the disappearance of hazardous waste oil from major ports in the country, the waste oil having been imported back into the country as lubricating oil. India Together (2006).

equally have been the consequence of a sustained campaign of opposition by environmental NGOs both in Europe and India and the potential political embarrassment to an impending state visit to India by the then President of France.

The *Clémenceau* was a former French aircraft carrier and flagship of the French navy. Decommissioned and laid up in 1997, the ship subsequently spent some four years being towed around the Mediterranean, seeking some refuge where pre-scrapping decontamination could be carried out. The final years of the vessel reflect the growing reluctance of European nations to accommodate so large a consignment of hazardous waste. Despite the large amount of hazardous material on board, the ship was particularly attractive to shipbreakers as it contained an estimated 22,000 tonnes of steel. Warships also tend to be plentiful sources of non-ferrous metals.

After being disarmed in October 1997, the *Clémenceau* was finally decommissioned and condemned in December 2002. Officially renamed *Shell Q790*,[38] ownership of the vessel passed from the hands of the military authorities to the National Directorate of Public Domain Interventions (DNID), the Ministry of Economy and Finance, at that time being responsible for the sale and demolition of old naval ships. A variety of uses were examined for the vessel; the first option, to sink her as a military target, was discarded. A second option, to sink her as an artificial reef off the coast of Marseilles, was also deemed unacceptable to the Environment Ministry, because of the high level of hazardous materials aboard. The decision to remove the asbestos and other toxic material was taken to comply with European waste regulations; the ship would then be offered for sale by public auction for demolition,[39] but remain the property of France until it was scrapped. After a bid from the French naval shipyard for the removal of the hazardous waste was deemed too expensive,[40] the contract was won by the Spanish company of Gijonese de Desquaces in April 2003. All hazardous waste was to be removed in Gijon on the Atlantic coast of Spain. The *Clémenceau* was towed from Toulon by an Italian company on 13 October 2003, but was spotted by a French patrol some 4 days later off the coast of Sicily,[41] *en route* to Turkey for breaking without the hazardous materials being removed. Since the contract had specified demolition in Europe, the ship was stopped by a French frigate (Turkey having refused entry to the *Clémenceau*), the contract with the Spanish company was cancelled, and France began negotiations with the second highest bidder, the specially created, Panama registered, Ship Decommissioning Industries (SDI)[42] for the removal of the visible

[38] This new official name was one that was almost universally ignored by all who reported its movements regularly, in favour of its traditional name (*Le*) *Clémenceau*.

[39] Zarach (2006).

[40] Orellana (2006).

[41] Zarach (2006).

[42] SDI was a specially created Panama registered subsidiary of the Eckhart Marine GmbH, itself a subsidiary of the ThyssenKrupp group. Eckhart Marine is a known cash buyer of end-of-life ships.

asbestos in Italy, but this venture failed due to cost overruns.[43] SDI then attempted to have the vessel decontaminated in Greece prior to despatch to Bangladesh for scrapping, but this attempt also failed, the ship being refused entry into Greece by the Greek military in November 2003. There were also reports that France had tried to send the ship to China, who had also refused entry.[44] Meanwhile, the *Clémenceau* lay at anchor off the coast of Sicily, until it was finally returned to the Navy yard at Toulon, where further, and unsuccessful, attempts were made by SDI, working with the French waste removal company Technopure, to decontaminate the ship fully. The plan was for the vessel then to leave for India, where the removal of the final asbestos and demolition of the hull was to be carried out at Alang by the Indian company Shree Ram.[45] On 22 December 2005, the Ministry of Defence announced the intention to export the ship to India, the ministry in charge of customs and the DNID formally authorising the export of 'war material.'[46]

In May 2005, it was reported that approval for the scrapping of the *Clémenceau* had been cleared by the Indian authorities,[47] although subsequent reports referred to the fact that a meeting of the Indian Supreme Court Monitoring Committee in February 2005 had laid down stringent conditions that must be met before the ship would be allowed entry.[48] Later in the year, in December 2005, the Administrative Court of Paris ruled to allow the ship to depart for India for scrapping,[49] the court claiming that the matter was beyond its jurisdiction, since the ship was deemed to remain a warship until it was completely scrapped[50] and hence beyond the provisions of the *Basel Convention*. Some 210 tonnes of asbestos had by this stage been removed, but this still left some 22 tons on board, plus a range of other toxic materials such as PCBs. The actual tonnage proved difficult to define accurately—22 tonnes is a figure that has been used by a number of sources. In 2006, Greenpeace cited large quantities of hazardous materials; between 500 and 1,000 tonnes of asbestos as well as other organic pollutants like TBTs, PCBs etc., the large figure presumably coming from the estimates later supplied by Technopure[51]

---

[43] Greenpeace (2005b).

[44] Kumar (2004). This report was a little more difficult to follow as China has made several attempts to obtain former aircraft carriers from various foreign navies.

[45] Under both French and European law, such an export of asbestos and other hazardous materials to China was forbidden.

[46] London (2006), p. 66. The Ministry's reasoning as to the ship being 'war material' made no difference. Since the ship had been disarmed, it no longer qualified as such.

[47] Ghatwai (2005).

[48] Greenpeace (2006).

[49] Ruling of the Administrative Court of Paris 30.12.2005.

[50] This was a convenient but questionable designation, since the ship had been largely de-equipped of operational material and had been used as a source of spares for other vessels, including a sister ship, the former aircraft carrier *Foch*, that had previously been sold by France to Brazil.

[51] Greenpeace (2006). The original estimate of the asbestos, made in 2003 when the ship was handed to SDIC, was at 215 tonnes—rounded to 200 tonnes as a safety margin. An early stage of the decommissioning revealed that the funnel lagging was made of fibre glass and not asbestos,

The former carrier left France for India on 31 December 2005, headed for Alang and thus began the second stage of her wanderings, the ship being detained at the Suez Canal by the Egyptian authorities for a week in January 2006 whilst evidence of compliance with international regulations was received from both France and from India.[52] However, on January 6, the Indian Supreme Court issued a split decision in their interim ruling that ordered the *Clémenceau* to stay out of Indian territorial waters whilst it sought further advice prior to making a final decision on February 13 as to whether or not the ship will be allowed to be broken there[53] once all relevant information had been submitted to the court. Whilst the Supreme Court was determined to keep the *Clémenceau* out of Indian territorial waters for the interim, the Indian government was eager to see the vessel demolished in India, not only for the employment and materials that it represented and to keep India at the forefront of the shipbreaking industry, but also to court favour with the French government.[54]

As far back as 2003, the Indian Supreme Court had laid down guidelines on ship recycling and the *Basel Convention*, which must be met, its order issued 14 October 2003 stating that it accepted the following recommendations of the report of the High Powered Committee on Management of Hazardous Wastes that:

> ...before a ship arrives at port it should have proper consent from the concerned authority or the State Maritime Board, stating that it does not contain any hazardous waste or radioactive substances...The ship should be properly decontaminated by the ship owner prior to the breaking, This should be ensured by the SPCBC.[55]

The subsequent conditions for the ship's entry were defined in 2 February 2005 by the SCMC sub-committee on shipbreaking and included:

- a report of the decontamination in Toulon and details of the actual quantities of asbestos and other hazardous materials removed from the ship;
- independent third-party audit verifying the report;

---

thereby removing a further 60 tonnes from the total to leave a revised estimate of some 160 tonnes present on the ship. Basel Action Network (2006).

Technopure subsequently advised that it had removed 155 tonnes, leaving 45 tonnes on board. However, the Defence Ministry announced that documentation from the landfill appointed to dispose of the hazardous waste accounted for only 85 tonnes, leaving the problem as to exactly how much material had been removed and how much remained on board. International Centre for Trade and Sustainable Development (2006).

Figures quoted had all been estimates, since the actual amount could not have been accurately quantified until the ship had been totally demolished. Nevertheless, the statement by the French Ministry of Defence on 12 February 2006 said that possibly through an administrative error, some 30 tonnes of asbestos had been misplaced. Zarach (2006).

[52] Egypt is also a signatory of the *Basel Convention*.

[53] Norohna (2006).

[54] The Indian military and national defence sector was eager to procure French weapons systems and France's recognition of India as a nuclear weapons power. In addition, the French president was to make a state visit to India on February 19 and 20th.

[55] *Supreme Court of India Civil Original Jurisdiction Writ Petition No. 657 of 1995* s70.2 (2), issued 14 October 2003.

– a certificate from the French authorities that the ship has been decontaminated and does not violate the provisions of the *Basel Convention*;
– the MoEF shall procure from the French Embassy all the relevant documents relating to the ship including an official statement from the French Government that, in its view and that of its own experts, the hazardous materials including asbestos have been removed up to 98 % and the balance would be recovered at Alang under guidance and co-ordination of the company (M/s SDI Ltd.) as well.

The Supreme Court's decision of February 13 came down in favour of constituting a new panel specifically to decide the issue, dismissing its Monitoring Committee, which had provided the Court with a report that came to no conclusive opinion.[56] Declaring that the SCMC was not sufficiently expert to decide on matters of toxic wastes,[57] the Court was to form a new committee of members nominated by the Ministry of Defence.[58]

Throughout the process, Greenpeace, together with BAN, Indian trades unions and local NGOs in both India and France had conducted its (by now very familiar) relentless campaign of lobbying and publicity against the despatch of yet another, but very significant, vessel to India for breaking. In its exasperation at all the attention that had been generated, the Indian Supreme Court took the extraordinary step of imposing a 5 day ban on all public discussion, demonstrations, protests and all further press reporting of the issue, claiming that they represented a challenge to the Court's authority; any transgressors would be held in contempt of court.[59]

Meanwhile, two summary proceedings had been put before the Paris Administrative Court by the NGOs seeking suspension of the decision to export the ship. In dismissing the claims,[60] the court stated that:

> none of the means put forward by the claimants at this state of the case are likely to raise any doubts about the legality of the orders under discussion.[61]

---

[56] Krishna (2006).

[57] This rather does put into question the role of the SCMC in other previous cases.

[58] Krishna (2006).

[59] Zarach (2006).

[60] The initial claim introduced by BAN and Andeva denouncing the contract to send the ship to India was declared admissible by the *Tribunal de Grande Instance* (TGI), the Civil Court of Paris in February 2005, but the following month it decided that it was not competent and that the claim should be reserved for the Administrative Court. After appeal, the Civil Appeal Court determined that the ship should stay in France for the time being and that the Civil Court was indeed competent to judge on the matter of asbestos decontamination; the claim could be reintroduced. A second decision of the French Civil Court of Paris in July 2005 again held that the claim was admissible but that the court was not competent to decide on the issue, which was again reserved to the Administrative Court, on the basis that the decision leading to the contract between the French state and the company SDI was 'an administrative decision' concerning the destination of war materials (as the ship was still described). This time the Appeal Civil Court in Paris upheld the ruling of the Civil Court.

[61] Two court orders dated 30 December 2005.

An appeal was therefore laid with the *Conseil d'État*[62] about the conflicting reports that had been issued concerning the amount of asbestos remaining on board the ship. The *Conseil* had to decide, firstly, whether the two orders were to be quashed and, secondly, whether the *Clémenceau* should be defined as waste, contrary to the argument of the French authorities that the *Clémenceau* remained a warship and therefore not bound by the *Basel Convention* or Regulation (EEC) 259/93,[63] amended 1 February 1993.

Referring to the precedence of Community Joined Cases *Vesso* and *Zaneth* and the *Zanetti* case (and subsequent cases),[64] which showed, according to the *Conseil*:

> ...the act of disposing of is not restrained solely to the abandonment of materials or substances involved and must be considered as waste material susceptible of being used for economic purposes, as long as they had not been regenerated or recycled and even if the holders had intention of selling them.

Directive 75/442/EEC (The Waste Directive)[65] defines waste as:

> ...any substance or object in the categories set out in Annex I which the holder discards or intends or is required to discard.

The fact that the State was disposing of the aircraft carrier by putting it up for public auction thereby qualified it as waste, even if the obsolete ship was partially intended for economic purposes.[66] The judgement of the *Arco Chemie Nederland and Others* (as well as *Palin Granit Oy*),[67] cited *The Waste Directive's* objective of protection, the ECJ reaffirming that:

> the third recital in the preamble to Directive 75/442 (The Waste Directive) states that the essential objective of all provisions relating to waste disposal must be the protection of human health and the environment against harmful effects caused by the collection, transport, treatment, storage and tipping of waste.

in addition to the Community policy of the precautionary principle and preventive action. These judgements were further reinforced by the opinion from *Paul van de Walle and Others*[68] that:

---

[62] European Commission, Association Ban Asbestos France and others, req. no. 288801, 288811.

[63] Article 2 states: '*For the purpose of this Regulation*; (*a*) *waste is as defined in article 1*(*a*) *of Directive 75/442/EEC (The Waste Directive), as amended by Directive 91/156/EEC*,' which defines waste as '*any substance or object in the categories set out in Annex 1 which the holder discards or intends or is required to discard*.'

[64] ECJ joined cases *Vesso and Zaneth* case C-206-207/88 28.3.1990, *Zanetti* case C-389/88 28.3.1990. [1990] ECR I-1461. Rulings on this subject have been regularly reassessed by the ECJ.

[65] *Directive 75/442/EEC, the Waste Framework Directive*, amended by *Directive 91/156/EEC*.

[66] Association of the Council of State and Supreme Administrative Jurisdiction of the European Union (2008). In addition the *Conseil d'État* followed ECJ case law—C-307/00 to C-311/00—according to which, in the case of a joint recovery and disposal operation, the more stringent scheme should apply *i.e*. disposal. Exporting waste to dispose of is strictly forbidden by Regulation 259/93, hence the export of the *Clemenceau*, which was of a mixed nature, was forbidden.

[67] In *Arco Chemie Nederland and others* Joined cases C-418/97 and C-419/97 [2000] ECR I-7411, and in *Palin Granit Oy* case C-9/00 [2002] ECR I-3533.

[68] *Paul van de Walle and others v Texaco Belgium SA* Case C-1/03.

> ...the decisive factor in determining the presence of waste is not assignment to a category of waste but rather whether the holder discards or intends to discard...

Regulation (EEC) 259/93 recognises two regimes for the shipment of waste; the first is shipment for recovery, for which a more flexible regime is provided, the second is shipment for disposal, which is very strictly controlled. Arguing that the shipment was for the purpose of recycling of steel and the disposal of asbestos, the French authorities sought the application of the first procedure. The *Conseil d'État* ruled that, where there was an option, the one most protective of the environment should be applied.[69] A further argument that the vessel was excluded from the scope of Annex II (the Green List) to the Regulation[70] as a:

> ...vessel...for breaking up, properly emptied of any cargo and materials arising from the operation of the vessel **which may have been classified as a dangerous substance or waste** (author's emphasis added)

was also dismissed, thereby rendering the vessel subject to Regulation (EEC) 259/93.

The *Conseil d'État's* rationale was based upon two different arguments, namely that:

- Category Q13 in Annex I of the Directive 75/442/EEC includes any materials, substances or products whose use has been banned by law. French decree of 24 December 1996 prohibits the use of asbestos fibres and of products containing such substances.
- The fact that the French Government had tendered a bid and entered into a sale agreement to decontaminate and demolish the ship established the State's intent to discard the hull of the 'former aircraft carrier Clemenceau.'

A written statement on behalf of the French Government had been issued in January 2006, aimed at justifying its position with regard to its intentions that the vessel be scrapped in India,[71] but on 15 February 2006 the *Conseil d'État* announced[72] that the ship was indeed waste, that its voyage to India would be suspended, and that the two summary orders enacted by the Paris Administrative Court on 30 December 2005 would be quashed. The ship's export documents were cancelled, the quantity of asbestos remaining on board rendering it subject to the *Basel Convention*. Since the vessel had been disarmed, the Ministry of Defence's argument that the *Clémenceau* was still considered to be 'war material' was deemed to be not relevant. Although the State was ordered to pay costs and damages to Greenpeace and BAN, the ruling did not call for the repatriation of the ship.[73]

---

[69] London (2006), p. 66.

[70] Regulation (EEC) 259/93 Annex II category GC030. A vessel for disposal which is 'properly emptied' may cease to become hazardous, but still remains waste.

[71] For a line-by-line critique of the statement, see Basel Action Network (2006).

[72] Order 15 February 2006, no. 288801, 288811.

[73] London (2006), p. 66. Declaring the vessel waste but not requiring its return was a routine that was to be repeated by the American EPA with regard to the *Oceanic/Platinum II*—see Sect. 5.8.

However, as President Chirac was about to make a state visit to India 4 days later, the potential for political embarrassment was high; he therefore took direct control of the ship from the ministries and ordered it back to Toulon.[74]

Whilst the result in France was one of considerable political fallout, the result in India was perhaps somewhat more devastating; with shipbreaking at Alang in steep decline and few businesses remaining in operation at the time. The loss of the *Clémenceau* meant the loss of a major unit. The ship may have taken considerable time to demolish, given the complexity of a major naval vessel in comparison to the relative simplicity of a merchant ship. Reports in the Indian press had placed a scrap value of US$15 million on the ship.[75] The yard in question was hoping for a large investment and the GMB announced that it had established a new waste reception facility for the treatment and disposal of wastes such as asbestos in 2005. Joint training for the workers at the Indian yard had been carried out with the French contractors, who were to have been on site in India to oversee the removal and disposal of the hazardous materials—this could have demonstrated a major development and model for the capabilities of an Indian operation in the style of that in force in certain of the Chinese shipbreaking sites.[76]

On 1 July 2008, the final demise of the *Clémenceau* was defined,[77] when it was announced that the ship was to be broken up at Hartlepool, by Able UK, the Teesside company at the centre of the ‘Ghost Ship’ saga, which began in 2003 with the import of four aged ships from the US Reserve Fleet. The former French carrier finally arrived at Hartlepool for scrapping on 8 February 2009.

The history of the attempts by the French Government to rid themselves of a former major asset, which had now become a major liability, illustrates a wide range of issues—the growing reluctance of European nations to become embroiled in such a contentious issue related to international law; the legal attempts by the authorities of authorities in both France and India to justify their actions; the disregard that was sometimes shown (both in this issue and others) for rulings from the Indian Supreme Court; and the possibility that was unfortunately lost to raise the standards at one Indian breaker’s site to those expected to obtain in Europe. It also demonstrated the lack of competence—self-confessed by the court

[74] The voyage back to Toulon was to be around the Cape of Good Hope, in order to avoid another passage via the Suez Canal (and more potential problems with Egypt). Zarach (2006).

[75] International Centre for Trade and Sustainable Development (2006).

[76] A close partnership was formed between SDI and Technopure on the one hand and the Indian breaker company Shree Ran Scrap Vessels and the waste management company, Luthra Group, on the other. The Indian operators had been trained in France by a certified training centre to deal with the asbestos removal and then trained on site with a SDIC sub-contractor Prestosid SAS on all the required equipment and technologies. Staff of SDI and Prestosid were also to be on site in India to monitor the dismantling. This was claimed to be the first time that such co-operation between an Asian shipbreaking company and European countries had been formulated. Zarach (2006).

[77] Lloyd’s List (2008).

in Paris, and by the sacking of its Monitoring Committee by the Supreme Court in India—for these bodies to address the problems associated with regulating the problems associated with hazardous wastes.

The ship's wanderings were initiated by economic reasons, namely the original announced cost of de-contamination by the Navy yard at Toulon, and subsequently by the refusal of other states to accept the vessel. Although details of most of the costs of the resultant peregrinations were not published,[78] with the benefit of hindsight this 'economy' would appear to have become a most expensive saving, given the costs that must have been involved in towing the dead ship around the Mediterranean, the despatch of the French navy to halt her progress to Turkey, the return to Toulon, the eventual tow to Alang (and back) and the final tow from the Mediterranean to Hartlepool.

5.5 Norway/Blue Lady 2005–2007

The case of the *France/Norway/Blue Lady* is that of a ship past its useful life and ready for disposal, another major asset becoming a major liability. The extended planning for disposal, aided by a crippling accident to the ship, included dubious stated intentions as to her future to facilitate her export, and a succession of owners between her departure from Miami and her final arrival at Indian breakers. These proceedings were aided by extended and sometimes conflicting and apparently spurious (but eventually successful) arguments accepted by the Indian courts in the face of conflicting, external expertise as well as by India's own interpretation of the *Basel Convention* as applied to ships-for-disposal and hazardous wastes. The ship's eventual entry into Indian waters was finally endorsed on 'humanitarian' rather than legal grounds and the stated relevance and significance of human and environmental protection eventually fell victim to the potential supply of resources and labour opportunities that the scrapping of the vessel represented.

Built in France and launched in 1960 as a ship that could operate as both a scheduled passenger liner and cruise ship, the *France* was probably the last of the true transatlantic liners, and at that time the longest passenger ship in the world. As the transatlantic sea trade gave way to air transport, high operating costs, and repeated labour strikes, resulted in the ship being laid up at Le Havre in 1975, where she lay for almost 4 years. In 1978, she was sold to Norwegian Cruise Line (NCL) and, after extensive reconstruction, re-emerged in 1980 as the *Norway*, the first of the big cruise ships, setting the standard for size and offerings. Her career as such was highly popular.

As newer ships began to enter the cruising fleet, the upkeep of the *Norway* began to suffer and she experienced various mechanical breakdowns, fires, and port

[78] An anecdotal comment set the cost of the tow to and from Alang alone at an estimated €3 million.

detentions for repairs. After the terrorist attacks in the US in 2001, plus an inadequate refit that failed to address the mechanical problems prevailing, the ship was put on cheap cruises from Miami, where on 29 May 2003 a boiler explosion in the ship left several crew members dead and the ship immobile. The *Norway* was subsequently towed to Bremerhaven, Germany, ostensibly for the costing of repairs. The following March she was declared surplus and transferred to NCL's parent company Star Cruises PLC Malaysia.

In May 2005, the ship was towed from Bremerhaven to Port Klang, Malaysia, the German authorities being advised that the ship was to be repaired in Asia for further operations in Australia, and there began the pattern of deception that is a feature of many of these earmarked vessels. In fact the ship was to be sold for scrap, and was visited by shipbreakers from India and Bangladesh whilst she lay off Port Klang, where her name was changed to *Blue Lady*. A report compiled by the NGO Shipbreaking Platform[79] highlighted statements made in the 2005 annual NCL report, submitted to the US Securities and Exchange Commission in 2006,[80] explaining how NCL had reduced the value of the ship by US$14.5 million during 2004, leaving the ship with a scrap value of just US$12.3 million,[81] a sum less than the estimated cost of €17 million needed to partially decontaminate the asbestos present on the vessel. The report also indicated that:

> NCL's management came to the conclusion that the sale of the vessel to interested third party buyers for reuse was not likely, signalling the company's resolve to dispose of the vessel by the end of 2004.[82]

The issue of the report coincided with the NGO Shipbreaking Platform's request to the Senator of Building, Environment and Transportation of Bremen for Germany's compliance with its obligations under *Basel* (Art. 9) and under the *WSR* (Art. 16) to take back the vessel for decontamination and to pursue its owners, Star Cruises Ltd., for illegal export.[83]

The ship then languished for some 5 months off Malaysia before departing for Bangladesh in January 2006, after her purchase had been announced by a Chittagong shipbreaker for the sum of more than US$12 million. However, a campaign of

[79] NGO Shipbreaking Platform (2006a).

[80] United State Securities and Exchange Commission Form 20-F. Annual report for the fiscal year ended December 31, 2005. Commission file number 333-128780 NCL Corporation Ltd, incorporated in Bermuda. Filed March 28, 2006.

[81] The NGO report cited NCL's 2005 annual report as determining the ship's value to be US$46.5 million, the face value of a non-interest bearing promissory note received by NCL from Star Cruises Ltd when the ship was transferred to the parent company. That US$46.5 million face value was made up of US$19.7 million proceeds from the insurance company following the 2003 boiler explosion, and US$26.8 million, the perceived value of the vessel upon sale. It also stated that in 2004, a reduction in the value of the Norway by over a half presumed that the ship would generate very little cash flow which would not cover the costs incurred—essentially the ship had been written off. NGO Shipbreaking Platform (2006a), p. 6.

[82] NGO Shipbreaking Platform (2006a), p. 2.

[83] NGO Shipbreaking Platform (2006b), p. 2.

protests[84] resulted in the Bangladesh government banning entry to the ship by inter-ministerial decree, the Environment and Forest Minister instructing the Central Bank and customs authorities not to allow the issue of an import order and the coastguards being ordered to keep the vessel out of territorial waters.[85] The sale was subsequently cancelled and the *Blue Lady* returned once more to Port Klang.

A scheme to have the ship reside as a luxury hotel/tourist centre/restaurants and museum in Dubai also did not come to fruition and the *Blue Lady* was next towed to Fujayrah in the UAE where, after a short stay, she was moved around for some time until, after various attempts at sailings to breakers' sites, the ship finally set off under tow for India, the Malaysian authorities having been informed that the ship was destined for Dubai and repairs.[86]

Prior to her arrival, however, an Intervention Application (IA29) had been filed by the Indian Platform on Shipbreaking with the Indian Supreme Court to ensure that the ship was not allowed entry to India in contravention of the Court's own directions of 14 October 2003. Thus began a campaign of protest and counter-protest involving environmental campaigners, the Supreme Court, the Supreme Court's Technical Experts' Committee on Management of Hazardous Wastes relating to Ship Breaking (TEC),[87] and the *Blue Lady's* new owners, Hariyana Ship Demolitions Pvt (HSD). The ensuing process was tortuous, contradictory in places, and incomplete in others, and whilst the following account is a summarised version of the proceedings, it serves to illustrate the difficulties in defining points of liability in the face of determined obfuscation. Following the application, the Supreme Court issued notices to the relevant authorities and the *Blue Lady* was barred from entering Indian territorial waters by the Gujarat Pollution Control Board. Representation was then made by HSD, requesting that the ship be allowed to enter the Alang anchorage, pleading the cost of holding the ship under tow to be some US$30,000 per day and giving assurances that should the TEC not allow the ship to be broken at Alang, they undertook to '*tow the ship away into international water, which can be done with ease*,' whilst adding subsequently that '*inspection of the ship can only be done at safe anchorage at Alang*.'[88]

Submissions made by opponents to the scrapping were next made to the TEC, reinforcing the need for observance of the Supreme Court Order of 14 October

[84] The ship had been included in the Greenpeace watch list of 50 ships.

[85] The minutes of the Inter-Ministerial meeting held on 15.02.06 detailed the decisions as—the Bangladesh Bank shall be requested to give necessary directions to the commercial banks not to open a letter of credit to import the ship; the Navy, Chittagong Port Authority and Coast Guard Authority shall be requested to take the necessary steps not to allow the entry of the ship into Bangladesh; the National Board of Revenue shall be requested to take necessary measures regarding non-entry of the ship into Bangladesh. An English version of the decision is included as Annex 1 to the NGO report, NGO Shipbreaking Platform (2006c).

[86] Specis and Tan (2008).

[87] This Committee was appointed by the Supreme Court in February 2006 to assist with the hearing on the *Clémenceau*, then in progress.

[88] Specis and Tan (2008).

2003, the *Hazardous Waste Rules 2003*, the *Factories Act*, and India's obligations under the *Basel* and *Stockholm Conventions*—obligations which had been firmly demonstrated by Denmark's action with regard to the *Riky* and Bangladesh's action with regard to the *Blue Lady* earlier. These constituted IA30, issued by the MoEF, which were in turn submitted to the Supreme Court, together with the interim report of the TEC. The relief sought under the report was for anchoring only. The submission also included letters from the owners, HSD, which, in some contradiction to earlier comments, stated that

> ...the currents in high seas in this part of the ocean are very strong especially when monsoon setting in [sic]. Therefore it would not be a safe location to anchor ship as the current may break the Anchor and the ship may start drifting.

HSD also pleaded human risks and problems regarding the 13 Indian crew members on board, a deliberate piece of timing, according to campaigners, aimed at coinciding with the onset of the monsoons after the ship had lain idle in Malaysia for 8 months.[89] The MoEF's application IA30 was disregarded by the Supreme Court, who on 5 June 2006—ironically Environment Day—gave HSD permission to anchor the *Blue Lady* off Alang on *humanitarian* grounds, since there were no *legal* grounds that could be applicable, none of the prior consent conditions being present from any of the states involved, no adequate provisions for the treatment of PCBs being available, etc. This permission did not include provisions for any demolition to commence. It did, however, violate the Supreme Court's own order of 2003. For some unknown reason, the response of HSD was immediately to tow the ship back to Fujayrah, UAE for 2 days, before returning to the edge of Indian territorial waters.

The next phase began with the inspection of the ship, a 6-member Committee being formed by the GPCB on 1 July 2006 to verify all hazardous materials on board and the assessment submitted by the Gujarat Environment Protection and Infrastructure Ltd. (GEPIL). A pre-assessment team first boarded the ship, at anchor some 25 nautical miles off the port of Pipavav, and the inspection was carried out between 8 and 10 July. The committee itself then boarded the vessel to verify the pre-inspection team's report and the GEPIL report, and the following day visited the Alang yards to inspect the removal, handling and disposal of ACM.[90] Following its inspections, the committee submitted its report to the GPCB, concluding that no hazardous material was observed in loose form except some oily rags; that no other hazardous material

> ...of any kind or quantity was found that cannot be safely removed, handled and disposed of at Alang; the Committee is thus of the opinion that safe removal, handling and disposal of asbestos, ACMs, PCBs and other hazardous waste can be done at Alang[91]

accepting thereby that there were indeed hazardous wastes on board.

[89] Mediavigil (2007).

[90] NGO Shipbreaking Platform (2006d).

[91] NGO Shipbreaking Platform (2006d).

The report was subject to an in-depth examination by the NGO Shipbreaking Platform,[92] who began by pointing out that the provision of a comprehensive inventory of on-board hazardous materials is the responsibility of the ship's owners, prior to the arrival of the ship in India. The conclusions of the Committee were deemed to be seriously flawed in several respects. The Committee's assertion that only 'cargo' can be included in the definition of waste was legally inappropriate. The belief that the inbuilt materials could be safely handled and disposed of in an environmentally sound manner was unsupported.[93] Both the pre-assessment and verification teams relied solely upon observations to determine the presence of hazardous wastes; asbestos contamination is often in the form of airborne microscopic particles, many of which had been released by the boiler explosion that occurred in Miami and the presence of PCBs can only be determined by laboratory analysis. The Committee also failed to have regard to two previous technical inspections of the ship, the first of which was undertaken in 2004 by Tecnitas, a French company employed by a prospective purchaser to conduct a qualitative and quantitative study of on-board asbestos, and who was given access to all of the ship's plans and documents. Their report outlined the areas where asbestos was present. The second study was undertaken by Ship Decommissioning Industries SAS (SDI), who inspected the ship in 2005 whilst it was docked in Bremerhaven; their report indicated airborne asbestos contamination over several decks, together with the presence of PCBs. The head of SDI had written to the MoEF expressing his concerns over the presence of airborne asbestos. Neither study was used by the Committee.

Of the documents finally considered with regard to permission for beaching, a report filed by the (hitherto unmentioned) Priya Blue Industries Pvt. Ltd. (PBI),[94] now claiming to be the owners, differed in its findings and conclusions of that of the MoEF. The TEC's own report was based on documentation supplied by HSD, which claimed no quantification of PCBs had been made by either the Inspection Team or by GEPIL. An approximation that had been completed indicated less than 10 tons. Accordingly, the TEC granted that the ship be given beaching permission, authorizing GMB so to do (subject to final approval of the Supreme Court), and on 3 August 2006 the GMB wrote to PBI granting permission for beaching according to the various provisions of the *GMB Ship Recycling Regulation 2003* and the directions of the Supreme Court previously referenced *and* after the submission of

---

[92] NGO Shipbreaking Platform (2006d).

[93] The NGO report refers also to a report that was submitted to them by Mr A. B. Andersen, of MetaFil AS, (a Norwegian company that focuses on developing maritime environmental technologies) who believed that once breaking operations begin, **the spread of contaminated materials can increase by a factor of 10 because of the methods employed in the Indian yards** (emphasis added), which fail to isolate these substances both during and after removal. This report was also submitted to the TEC, 26 June 2006.

[94] The role of PBI was hitherto never defined, nor the date when its ownership became effective. On 1 August 2006, the *Business Standard* reported that HSD had sold the ship to PBI for US$16 million. Specis and Tan (2008).

all required documentation by PBI. The *Blue Lady* was beached off Alang, some 4,000 ft. offshore[95] on August 15, no documentation being submitted. The final report of the TEC was submitted at the end of August 2006 and was signed only by the Chairman.[96]

The question of whether authorization for dismantling could go ahead came before the Supreme Court in early September 2007 under Intervention Application 34 of 2006.[97] The Court's judgement allowing the scrapping of the vessel was issued 10 September 2007.[98] In comparison to other judgements examined in this chapter, this court report is a descriptive and somewhat rambling examination, beginning with a description of the ship's 'golden period' and continuing not so much with the legalities of hazardous materials and *Basel* obligations, but with rather more of a discussion of environmental *versus* economic considerations and emerges more as a commercial justification for scrapping.

The Court accepted the reports compiled by the various inspection bodies whilst making no reference to the technical reports submitted by outside specialists. Information provided by a former project manager of the ship referred to radioactive material in some 5,500 places in the ship's fire detection system whilst the TEC report merely stated that 12 smoke detectors had been collected and '*handed over to the breaker for keeping safe and securely pending their disposal at BARC through AERB*.'[99] The statement made by the TEC that, since the vessel was now grounded, this was an irreversible process and the ship could no longer be refloated, hence scrapping appeared to be the only option,[100] was also refuted by more than one salvage company. In a letter to the Prime Minister, Allen Busch, the Vice President and General Manager of Titan Salvage stated that his firm had both the capability and expertise to refloat the vessel and that there was 'a very high probability that the *Blue Lady* is not at all in "*an irreversible position*" as the esteemed court has found.' This position was also repeated by the firm of Aage Anderson, which had earlier been involved in the *Clémenceau* case.[101] These issues were also not included in the Court's judgement.

---

[95] This long distance was due to the fact that this huge vessel had arrived as a dead tow, unable to beach under its own power.

[96] It was customary for such Final Reports to be signed by the whole committee. This was a practice that was also followed by the Supreme Court Monitoring Committee on Hazardous Wastes and the Supreme Court Committee on Waste to Energy. Specis and Tan (2008).

[97] Citing Writ Petition (civil) 657 of 1995, the petitioner being the Research Foundation for Science, Technology and Natural Resource Policy, the respondent being the Union of India and others.

[98] Supreme Court Judgement Order IA No. 34 of 2006 in Writ Petition (civil) No. 657 of 1995. S.H. Kapadia, J.

[99] As cited in Mines and Communities (2007). This article also states that the Supreme Court inquired as to whether India was signatory to *Basel Convention*—which, if correct, is a most curious admission for an apex court.

[100] A report requested by the Supreme Court from the TEC and submitted on 10 May 2010 concluded that '*beaching was an irreversible process*.'

[101] Krishna (2007).

Removable capabilities of the shipbreakers with regard to asbestos were characterised by the TEC as 'State-of-the Art;' however, the fact that some 80 % (later 85 %) of the asbestos content was deemed to exist in ceiling and wall panels, which were re-usable, was mentioned some five times in the judgement, demonstrating again the attractiveness of asbestos as a marketable commodity in the Indian market, rather than as a hazardous waste for disposal.

In substantiating its decision, the Court made several references to the principle of balance and proportionality, citing a Keynote Address by Lord Goldsmith, Her Majesty's Attorney General (UK),[102] who stated that the unwritten British Constitution is based on the three principles of rule of law, commitment to fundamental freedoms, and the principle of proportionality. The European Convention of Human Rights (ECHR) and its concept of balance were also cited. Judgement on the *Blue Lady* was based on a number of precedents, including the case of the *Research Foundation for Science Technology v. Union of India* (2005),[103] where it was held that the precautionary principle is a part of the concept of sustainable development. At paragraph 16 of that judgement, reference was made to *Vellore Citizens' Welfare Forum v. Union of India*,[104] in which it was held that:

> ...the precautionary principle and polluter-pays principle have now emerged and govern the law in our country, as is clear from Articles 47, 48-A, and 51-A(g) of our Constitution and that, in fact, in the various environmental statutes including the Environment (Protection) Act 1986, these concepts are already implied. Further it was observed in the *Vellore Citizens' Welfare Forum* case that these principles have been held to have become part of the customary international law and hence there should be no difficulty in accepting them as part of our domestic law.

The application of these principles was further enforced by reference to the case of the *A.P. Pollution Control Board v. Prof. M.V. Nayudu*,[105] in which it had been further observed that:

> The principle of good governance is an accepted principle of international and domestic laws. It comprises of the rule of law, effective State institutions, transparency and accountability and public affairs, respect for human rights and the meaningful participation of citizens in the political process of their countries and the decisions affecting their lives.

The *Research Foundation for Science Technology* judgement continued with further references,[106] concluding that:

[102] Stanford (2007).

[103] *Research Foundation for Science Technology v. Union of India* (2005) 10 SCC 510.

[104] *Vellore Citizens' Welfare Forum v. Union of India* (1996) 5 SCC 647.

[105] *A.P. Pollution Control Board v. Prof. M.V. Nayudu* (1999) 2 SCC 718.

[106] Reference was also made in para. 16 to Article 7 of the draft approved by the Working Group of the International Law Commission in 1996 on 'Prevention of Transboundary Damage from Hazardous Activities' to include the need for the State to take necessary 'legislative, administrative and other actions' to implement the duty of prevention of environmental harm, and also to *People's Union for Civil Liberties v. Union of India* (1997) 3 SCC 433.

Environmental concerns have been placed on the same pedestal as human rights concerns, both being traced to Article 21 of the Constitution and The Basel Convention, it cannot be doubted, effectuates the fundamental rights guaranteed under Article 21.

At paragraph 43 of the 2005 judgement, the Court also stated that '*We are not suggesting discontinuing of ship-breaking activity* [*sic*] *but it deserves to be strictly and properly regulated*,' the authorities needing to be vigilant about the generation of hazardous waste if the appropriate actions by the various agencies, especially the Maritime Board and the SPCB, are not taken.

Returning to the concepts of balance and proportionality *versus* sustainability, the Supreme Court then cited *T.N. Godavarman Thirumalpad v. Union of India and Ors.*,[107] in which it was stated that:

It cannot be disputed that no development is possible without some adverse effect on the ecology and environment, and the projects of public utility cannot be abandoned and it is necessary to adjust the interest of the people as well as necessary to maintain the environment. A balance has to be struck between the two interests. Where the commercial venture...would bring in results which are far more useful for the people,[108] difficulty of a small number of people has to be bypassed...the convenience and benefit to a larger section of the people has to get primacy over the comparative lesser hardship.

This effectively represents an argument that might be applied to any and all ships arriving for breaking. The apparent selectivity of the Court and the reports it considered does not, however, demonstrate an impartial approach to 'balance.' Since the question of the refloating of the ship was also bypassed as being 'irreversible,' on 11 September 2007 the Court granted permission to the shipbreaker to dismantle the *Blue Lady*. Stripping the ship began in October, demolition in the following January, and by late 2008 demolition had been completed.

5.6 Ben Ansar/Beni/Aqaba Express/Al Arabia 2007

In what had by now become almost a routine event, the *Beni Ansar* represented a ship sailing for disposal without due authority, and to a spurious destination and future. The frequency with which this vessel underwent changes of identity must be a measure of the urgency with which its owners sought to hide a ship identified as 'problematic.'

In 2007, despite assurances made to the contrary to the Spanish authorities, the former ferry sailed from Almeira to Alang for scrapping. Under advice from the European Commission an arrest warrant had been served on the ship, which was operating under a certificate stating that it was on a final voyage to Alang or Chittagong for scrapping (having actually been sold to Indian breakers).[109] The

[107] *T.N. Godavarman Thirumalpad v. Union of India and Ors.* (2002) 10 SCC 606.

[108] The Court stated that scrapping the *Blue Lady* would employ 700 workers and provide 41,000 metric tons of steel, thereby putting less pressure on mining activity elsewhere.

[109] Mendez (2007).

vessel was detained on July 12 and a detailed inspection indicated the presence of hazardous materials on board. During that month the ship changed its name and registration with different open registers, from *Beni Ansar* under the Morroccan flag, to *Beni* under the flag of Tuvalu, and then to *Aqaba Express* under the flag of the Comoros Islands.[110] The ship was finally allowed by the authorities to sail from Almeira ostensibly to Constanza in Romania on September 7, for the purpose of refurbishment.[111] At sea, the ship then changed course, arriving on October 27 to anchor off Alang under yet another name of *Al Arabia*. This deceptive departure classified the vessel as illegal traffic under the *Basel Convention*. The vessel also failed to comply with the Indian Supreme Court's orders of 14 October 2003 and 6 September 2007.[112] Representations from both the Spanish Ministry of Public Works and Economy and from the UN's Basel Secretariat to the Indian authorities failed to stop the scrapping of the vessel. According to the European Commission:

> Spain as the presumptive country of export is in the best position to take further measures, i.e. giving full information about the hazardous nature of the waste, requesting return of the ship and/or offering cooperation and environmentally sound management elsewhere, and initiating criminal procedures against the persons who have given false declarations towards Spanish authorities (in so far as this is criminal under Spanish law). The EU or Commission can only play a limited role in this context.[113]

5.7 Otapan 2006–2008

The *Otapan* case represents the long process, drawn out over a number of years by the owners, to legitimise the scrapping of the ship. This case bore many similarities to the process to which the *Sandrien* was subjected, both ships being obsolete chemical tankers that were eventually decontaminated at the expense of the Dutch authorities. It illustrates the ease with which its owners evaded not only liability for its controlled disposal, but also for the cost of that disposal.

The *Otapan* was built in 1965 and owned by a Mexican shipping company Mavimin who, in 1999, put the ship up for disposal whilst it was lying in Amsterdam. In 2001, members of the crew were caught illegally trying to remove asbestos from the vessel, which was eventually seized by the Dutch authorities. In 2005, the vessel was sold to Basilisk, who gave notice of their intention to remove the ship to Turkey. In July 2006 an export permit was granted by the Secretary of State[114] to

---

[110] Since its completion in 1975 the ship had had several names prior to *Ben Ansar*, including *Wisteria* and *Princess Maria Emerald*.

[111] A repeat of the intended plan for *Sandrien*—see Sect. 5.1.

[112] Ban Asbestos Network of India (2007).

[113] As quoted in Ban Asbestos Network of India (2007).

[114] An earlier denial of sailing permission had been issued by the Minister on the grounds that no financial guarantee had been posted by Basilisk. This was rescinded a week later when the said bank guarantee was provided.

allow the ship to sail to Turkey for scrapping by its new owners the Şimşekler Ship Recycling Company of Aliağa. However, whilst the ship was en route to Turkey, investigations by Greenpeace Netherlands discovered that the inventory of hazardous waste on board, and hence the export licence, were incorrect. The Turkish authorities were advised of this situation by the NGO Shipbreaking Platform (of which Greenpeace is a member), and permission to allow entry to the ship was consequently refused.[115] There then began a series of visits by the Dutch authorities to Turkey and the Turkish owners to the Netherlands; meanwhile, the *Otapan* was returned to Amsterdam in October 2006 by the Turkish Ministry of Environment, who were insisting on a full pre-cleaning.

A case against the Secretary of State's decision to allow the vessel to sail to Turkey was submitted to the Council of State, Administrative Jurisdiction Division, naming Stichting Greenpeace Nederland and others as claimants and the State Secretary for Housing, Spatial Planning and Environment as defendant.[116] The defence began with a claim that the claimants could not be regarded as 'interested parties' whose interest was directly affected by the order as defined under section 1.2 of the *General Administrative Law Act*. It was determined that under the provisions of their bylaws, Greenpeace's object of promoting nature conservation by ending certain abuses were affected by the Minister's order and that consequently the Foundation did have the status of an interested party. Such status was not granted to the other parties, including the Turkish trade union Limter-Ẏḅ.

The case then moved to a consideration by the court to determine whether the Minister's order to permit the sailing was correct. The charge by the claimants that Basilisk could not be treated as notifier[117] of the waste since it could not be treated as the original producer or holder of the waste was rejected by the Council, but the claim that the defendant had wrongly failed to classify the scrapping of the ship as a disposal activity was upheld. Reference was made by the court to previous judgements of the ECJ regarding the validity of classification of waste handling processes, beginning with the *SITA EcoService Nederland* (2003) case,[118] whereby when a waste treatment process comprises several stages, it must be classified as either a disposal or recovery operation, taking into account only the first operation that the waste has to undergo after shipment. Reference was also made to *Abfall Service AG* (2002)[119] and *Oliehandel Koeweit BV and others* (2003) with joined

[115] Basel Action Network (2008a).

[116] Council of State, The Hague. *Stitching Greenpeace Nederland v the Minister of Housing, Spatial Planning and the Environment* (*Decision on the Otapan*) Case No. 200606331/1 211.2.07.

[117] As per Article 2 of the Council Regulation (EEC) 259/93.

[118] *SITA EcoService Nederland, formerly Verol Recycling Limburg BV v Minister van Volkshuisvesting, Ruimtelijke Ordening en Milieubeheer* Case 116/01 [2003] concerning the competence of authorisation of despatch to scrutinise the classification of the purpose of a shipment. This, and the following cases (below), were concerning the interpretation of *Regulation (EEC) 259/93* and *Directive 75/442/EEC*.

[119] *Abfall Service AG (ASA) v Bundesminister für Umwelt, Jugend und Familie* Case no. 6/01 [2003].

cases,[120] which determined that the test of whether the operation was either recovery or disposal was to be determined on a case-by-case basis, depending on whether the purpose of the operation was useful in that it replaced the use of other materials—in which case it would constitute recovery. From the permit to sail and the scrapping plan, it was determined that the scrapping consisted of several stages, no actual demolition being allowed until all the asbestos had been removed. Despite the fact that scrap removed from the ship was to be sent to the metalworking industry for recycling, the disposal was basically one of disposal and not recovery, hence the Minister's classification (and therefore the permission to sail) was incorrect and the plaintiff's case was allowed.

Combined pressure on the Dutch Ministry of Environment from the NGO Shipbreaking Platform, the Turkish Prevention of Hazardous Shipbreaking Initiative and Greenpeace Netherlands led to a formal inventory of on-board hazardous materials being compiled. This was followed in 2007 by a second clean-up, paid for by the Netherlands at a cost of €4 million when the Turkish owners filed for bankruptcy.

During the decontamination, some 411 tonnes of asbestos and asbestos-contaminated material was removed from the ship, leaving 850 kg of asbestos and 331 tonnes of bunker oil, which could not be removed without dismantling the ship and thereby rendering her unable to sail. That year, the Turkish yard also received permission to receive the ship for recycling, and the *Otapan* finally sailed from Amsterdam in May 2008, to arrive in Aliağa the following month.

The outcome of this protracted action resulted in the shipbreaker agreeing to carry out the demolition in accordance with the *Basel Guidelines on Ship Dismantling*. An agreement was made between the Dutch and Turkish authorities to look for a joint solution should further problems arise during the removal of the remaining asbestos. The NGO Shipbreaking Platform cited the events as an example for other European States to follow with regard to making shipowners liable for the breaking of their ships and produced their '*Otapan Principles*,'[121] which involved the pre-cleaning of ships prior to final voyage; compliance with the *Basel Convention*; and the environmentally sound management of the recycling and downstream waste operations, based upon the ILO, IMO, Basel Convention, and NGO Platform Green Ship Recycling standards.[122]

[120] *Oliehandel Koeweit BV and others, Case C-307/00*; *Slibverwerking Noord-Brabant NV, Glückauf Sondershausen Entwicklungs- und Sicherrungsgesellschaft mbH, case C-308/00*; *PPG Industries Fiber Glass BV, case C-309/00*; *Stork Veco BV, case C-310/00*; *Sturing Afvalverwijdering Nord-Brabant NV, Afverbranding Zuid Nederland NV, Mineralplus Gesellschaft für Mineralstoffaufbereitung und Verwertung GmbH, formerly UTR Umwelt Gmbh v. Minister van Volkshuisvesting, Ruimtelijke Ordening en Milieubeheer, case C-311/00* [2003].

[121] Basel Action Network (2008b).

[122] Basel Action Network (2008b).

5.8 Oceanic/Platinum II 2008–2010

The case of the *Platinum II* involves frequent changes of ownership and identity (and hence liability), allegedly falsified registration details, and another case of convenient circumstances at the intended point of demolition that might add weight to the urgency of the need to permit entry and, ultimately, scrapping. The case demonstrates that, although the US agency imposed financial penalties upon the exporting owners for contravention of environmental regulations, it was unable or unwilling to actually seek the return of the vessel to the USA.[123] It further demonstrates the ease with which ship owners can circumvent restrictions on the illegal export of ships for demolition by sale to intermediary parties; a false claim of registration; the way in which international and national regulations are interpreted in terms of definitions of waste; and the execution of prescribed inspections in a way that is both questionable yet profitable to the commercial parties involved. In this instance, the case also initiated an approach to inconvenient regulation in India that was centred upon a change to regulation rather than to the circumstances of the venture. This time, however, a somewhat more intransigent (Indian) authority response resulted in the vessel being wrecked whilst still some miles offshore. All these issues equate to a convenient sidestepping of liabilities.

The ocean liner *Platinum II* was originally built for the American Export Line as the SS *Independence* in 1951. In 1959, the ship was converted to a cruise ship and acquired the first of her new names *Oceanic*.[124] Since her completion, she changed her original name a further five times, her flag registration twice, and operated under a series of different owners before her final sale for demolition. This frequent change of identity is common and lawful practice as a vessel changes ownership during her operational life, but also shows the ease with which a vessel's identity and registration may be altered, sometimes in the attempt to hide the identity of the actual owners (although it is not implied that such was the case in this regard). In 2001, the ship was laid up in San Francisco following the bankruptcy of her then owners.[125]

In 2008, the *Oceanic* was finally purchased by cash buyers Global Marketing Systems (GMS) and was towed, ostensibly to Singapore—later changed to Dubai—

[123] Following a campaign against the demolition of the former *Exxon Valdez* at Alang (see Sect. 5.11), further concerns were expressed about the likely arrival of the former US ship *Delaware Trader* at Alang. Clearance was authorised by MARAD in June 2012 but it was reported that MARAD advised the USEPA of likely problems of PCB contamination and a failure to comply with the US Toxic Substances Control Act; the USEPA nevertheless allowed the export to proceed. The vessel actually arrived for demolition at Gadani Beach, Pakistan in early August 2012. Culture Change (2012).

[124] Between 1974 and 2009, the ship's name was changed from *Independence* to *Oceanic Independence* (1974), *Sea Luck I* (1974), *Oceanic Independence* (1975), *Independence* (1982), *Oceanic* (2006). Ownership was changed some nine times between 1951 and 2008.

[125] Details of the ship's information as listed in Lloyds Register during her laid up period cites the registered owner as California Manufacturing, but her beneficial owner as 'unknown.'

but rumoured actually to be heading for India or Bangladesh for scrapping. Warnings filed by BAN with the United States Environmental Protection Agency, advising that the ship was likely to be carrying PCBs and heading for scrapping in South Asia,[126] resulted in a fine imposed by the Agency in January 2009 of over US$½ million on GMS and Global Shipping, LLC (GSL),[127] an affiliate of GM, for exporting the ship without prior removal of hazardous materials,[128] this action constituting the illegal export of PCBs for disposal and use in commerce, contrary to the US *Toxic Substances Control Act* 1976 (TSCA).[129] GMS had to certify that they would not undertake such actions again. The owners, however, were not actually required by the Agency to return the ship to the US for removal of those materials, the EPA claiming that it had no authority to have the ship recalled.[130] Although the sale of US flagged ships to overseas breakers was prohibited by Vice President Gore in 1998, privately owned US vessels have been able to depart without hindrance simply by reflagging or registering to a new, foreign owner. MARAD, however, now has to advise the EPA of ships applying for re-flagging or scrapping, the latter's approval being required before reflagging can take place.[131]

In 2009, after having spent some months at anchor in Dubai, the ship, now renamed *DV Platinum II*, left Dubai for Alang under tow.[132] The ship was permitted to anchor on 7 October, after which inspections by the Gujarat Pollution Control Board (GPCB) and the GMB determined that although the vessel did contain

[126] Basel Action Network (2009).

[127] Fines comprising US$32,500 against GMS and US$486,000 against GSL.

[128] *United States Environmental Protection Agency, Region 9, San Francisco, Consent Agreement and Final Order Docket no. TSCA-09-2008-0003 relating to violation of 40C.F.R.§761.97 and section 15(1) of TSCA, 15 U.S.C.§2614(1) by exporting the OCEANIC containing PCBs and PCB items for disposal outside the United States.* The charges were not contested.

[129] Once charges were laid by the EPA, the owners of the vessel initially claimed that the ship was not going to be scrapped but reused as a vessel by her new owners. BAN subsequently claimed that while the EPA was taking legal action against the owners, the Maritime Administration (MARAD) was actually approving the sale of the vessel to a foreign buyer, offering support to Platinum Investment Services Corp. (PISC), based in Monrovia, Liberia. PISC '*appears to be a "mailbox company": under Liberian law, a company may register without publicly revealing an address, any principal owners, board members or spokespersons of any kind. The company has no office, no website and has no known history of ship operations. It is likely MARAD's authorization of the sale of the ship hampered the EPA's own efforts to demand the ship be returned for proper testing and remediation.*' There would appear to be little coordination between the two agencies. Basel Action Network (2009).

[130] SSMaritime (2009).

[131] Ramachandran (2009).

[132] Agreement of sale (dated 1st October 2009) for the ship lying 'as is' by its then owners Platinum Investment Services Corp. Monrovia, listed the buyers as M/S Hariyana Ship Demolition Pvt. Ltd, Bhavnagar, the selling price being US$5 million. The agreement does not include details of the ship's registration, which was found to be falsified on arrival at Alang. Details of the buyer were subsequently changed to M/s Leela Ship Recycling Pvt. Ltd., Alang, on 12 October 2009, after beaching permission from the GMB had been authorised by the GPCB on 6.10.2009. The new sale price of the ship was listed as US$4.851 million.***

asbestos, there was no loose material on board; pollution clearance was therefore granted.[133] This inspection report, issued 10 October, cited more than 200 tonnes of asbestos-containing material and 20 tonnes of PVC cables and other PCB containing material on board. The inspection report of the Atomic Energy Regulatory Board identified the presence of some 653 sources (smoke detectors) in the accommodation areas of the ship, but '*as the vessel is dead, there is no crew member on board, all the smoke detectors will be removed by the buyer of the vessel.*'[134] However, following complaints filed with the Ministry of Environment and Forests with regard to toxic and radioactive materials on board the ship by the Indian Platform on Ship Breaking (IPOS), a Central Technical Team[135] was constituted by the Ministry to further inspect and report on the ship and this was carried out on 20 October 2009.[136] One of the complaints was that the ship was actually a warship, and hence a Special Category ship requiring as per Supreme Court Order quantification of hazardous wastes and substances well in advance of the arrival of the ship to facilitate desk review before anchoring permission could be granted. Subsequent requirements for quantification of hazardous waste were deemed to be unnecessary by the GPCB/GMB, who had categorized the ship as being only '*of general concern.*' However the terms of reference of the Central Technical Team included such an assessment on the quality and quantity of hazardous waste items, *based on the various documents submitted by the GPC and other agencies*. This complaint by the IPSB that the *Platinum II* was actually a warship was unfounded, the ship having all the characteristics and obvious appearance of a large ocean-going passenger ship. Furthermore, its identity could also have been easily confirmed by the unique registration number as issued by the IMO.[137] The report of 23 October confirmed that a desk review had been undertaken by the GPCB before recommending anchoring permission.

Included in the recommendations of the team was the fact that the category 'larger' passenger ship should be qualified to avoid further confusion on 'general concern' or 'special concern' ships and the relative approach to breaking that each requires. The report from the second inspection again quantified the extent of the on-board asbestos, but stated that it was difficult to carry out an assessment to quantify the amount of PCBs. The *Times of India* reported that this clearly put the import of the ship in violation of the *Stockholm Convention*, which bans the

[133] An earlier report by the surveyor on behalf of the M/S Pinnacle Marine Services Pvt. Ltd., Bhavnagar, found the ship to be 'portworthy' *i.e.* suitable to remain afloat at Alang Anchorage, in fair weather, subject to the vessel being suitably manned.

[134] Copies of the ship's papers, inspection reports and correspondence and a supposed Kiribati certificate of registration are all available online. Ministry of Environment and Forests (2009).

[135] The team consisted of Technical Officials from the Ministry of Steel, the Central Pollution Control Board and the Atomic Energy and Regulatory Board. Central Technical Team (India) (2009).

[136] Ministry of Environment and Forests (2009).

[137] Provision of the ship's IMO number is part of the information required to be submitted for anchoring permission.

movement of POPs. Further, India does not possess adequate facilities to treat or dispose of such hazardous material.[138] The sales agreement stated that the vessel be delivered free of cargo, hence the hazardous materials identified as being on board by the inspection teams posed a query as to their classification.[139] Since no prior informed consent to the import was produced by India beforehand, arrival of the ship would appear to be in contravention of the *Basel Convention* (to which India is a Party) and in violation of the Supreme Court's orders of 14 October 2003 and 6 September 2007, which require ships arriving for scrapping to be pre-cleaned of hazardous materials prior to import.[140] The owner of the *Platinum II* was reported to be Rajiv Reniwal of the Haryana Ship Breakers, the man who earlier acquired the *Blue Lady*.

Permission for the SS *Platinum-II* to be beached was then denied when it was discovered that her papers identifying her as registered in the Republic of Kiribati (in the name of M/s Platinum Investment Services Corporation, Monrovia)[141] had been falsified in order to evade the regulations on toxic materials; the claimed registration was rejected by the Kiribati registry.[142] A copy of the USEAP order was also supplied to the Ministry by IPOS and for both of these reasons—the violation of the *US Toxic Substances Control Act* and the false registration—the Ministry of Environment and Forests directed the GMB to disallow beaching and scrapping of the vessel at Alang, citing the precautionary principle behind their decision.[143] Whilst arguments concerning demands for the ship to be returned to the US were being aired,[144] the dead hull of the *Platinum II* grounded at Gopnath, south of Alang. There then followed reports from her owners to the Gujarat Maritime Board that cracks were appearing in the ship's hull and that final beaching at Alang

[138] Sethi (2009).

[139] A report issued by FutureNet Group, Inc., on the limited (sic) pre-decommissioning asbestos inspection undertaken 26–28 January 2008, whilst the ship was docked in San Francisco, indicated that some 147.3 metric tonnes of asbestos-containing material needed to be removed from the vessel before its demolition. This inspection was undertaken in accordance with the requirements of the United States Environmental Protection Agency 40 CFR Part 61 of the Federal National Emission Standards for Hazardous Air Pollutants, prior to any planned de-commissioning activities.

[140] Basel Action Network (2009).

[141] Letter dated 3.10.2009 from the Platinum Investment Services agent Compass Shipping Agency, Bhavnagar, to the GMB, GPCB and the Assistant Commissioner of Customs defines the ship's flag as Kiribati. The same letter requests that on arrival (due the following day 4.10.2009), '*looking to rough weather and heavy current at Bhavnagar Anchorage and considering the safety of the vessel immediately anchorage position is essential therefore you are requested to grant anchorage permission/position in view of Hon'ble Supreme Court.*'

[142] IMOWatch (2009). The tug Barracuda was also reported as being falsely registered in Kirabati.

[143] Financial losses from the cancelled breaking were announced by the owner of the Leela yard as being Rs 40 crore for the breaker, and Rs 4 crore for the government in Customs duty and Rs 1 crore sales tax. Pandi (2009).

[144] Indian Platform on Ship Breaking (2009).

was therefore essential to prevent the ship from becoming a total loss *in situ*. The leaks were sealed under the supervision of the Central Technical Team.

The importer's agent was asked by the GMB to remove the ship before the onset of the monsoon since the vessel might cause an obstruction to traffic for the port in the event of a storm or cyclone. The agent[145] then declared the ship a wreck which could not be refloated. It was next reported that the government demanded from the GMB an explanation as to why the ship was originally allowed entry, despite the Supreme Court's ruling that toxic materials should be offloaded at the point of departure before a vessel leaves for India. At this point, the problem now becomes one of definition, since the ship has been designated as a wreck, which leaves it no longer a viable vessel.[146] The response of environmentalists was to impose pressure to have all vessels entering Indian waters to state the purpose of their visit prior to anchoring, rather than simply arrive and be grounded. The response of the Union Ministry of Shipping was to propose a change in India's Merchant Shipping Act 1958 classifying ships aground as wrecks.[147]

Pictures of the ship published in May 2010 show the vessel with her back clearly broken just aft of the superstructure and down by the starboard bow. The vessel was demolished in situ, the wreck having been cleared by January 2011 and presumably at some big financial loss to the breaker once outstanding legal issues are settled. The potential for marine pollution by a vessel being broken up at sea is presumably of a far higher level than that of a vessel being broken up on the beach.

5.9 Onyx/Kaptain Boris 2009–2010

Although some major successes have been achieved by the NGOs, such as the extensive campaign against the *Clémenceau*, not all attempts have proven successful. The case of the former ferry *Onyx* displays elements similar to those of the *Riky*, but here it was the exporting state that failed to take suitable action rather than the importing state as was the case with the *Riky*.

Built in 1966, the *Onyx*[148] operated as a car and passenger ferry between Finland and Sweden for a number of years with numerous periods of lay-up and charter; she also appeared to be somewhat of an accident-prone vessel. In July 2007, the vessel, now renamed *Cassino Express*, was reportedly sold for scrap, subsequently changed to new ownership in St. Vincent and Grenadines for trading in the Caribbean area.

[145] Some press reports assign this declaration to the local authorities.

[146] IMOWatch (2010).

[147] Indian law relating to wrecks is covered by the *Merchant Shipping Act* 1958, Part XIII Section 2 (58), which defines a wreck as a vessel which has been abandoned *without hope or intention of recovery*—this is not the case of the *Platinum-II*. Krishna (2010a).

[148] Originally built as the *Fennia* in 1966, the ship became the *Cassino Express* in 2001, the *C. Express* 2007, the *Onyx* in 2009 and the *Kaptain Boris* in 2010.

Shortly afterwards, the vessel was prohibited from leaving Vaasa, where she was laid up, after the Finnish Environmental Institute SYKE announced a transport ban on account of the presence of asbestos, PCBs and other toxic materials found on board and the ship was declared to be waste. The ban remained in place pending a declaration of future use or an application to be made for a waste export permit.[149]

In July 2009, the export ban was lifted by SYKE after assurances from the then owner Attar Construction that the vessel was to be repaired and chartered to the International Shipping Bureau in Dubai. A letter from the NGO Shipbreaking Platform to the Finnish Minister of the Environment alerting her to the likely contravention of the *WSR* did not produce a positive response.[150] The vessel left Finland in November 2010, ostensibly heading for repairs in a Turkish shipyard—a claim denied by the shipyard in question. After a stopover in England, a machinery failure resulted in the ship being towed to Brest in France, where she was detained after the crew claimed that they had not been paid for 2 months.[151] The destination was then changed to Piraeus, Greece and later to Limassol, Cyprus. The ship eventually arrived at Dubai, in April 2010 where she was sold to Red Line Shipping Ltd., renamed *Kaptain Boris* and reflagged to Sierra Leone. The following month, the ship arrived at Gadani Beach, Pakistan, for scrapping.

5.10 Margaret Hill/Chill 2010

The case of the *Margaret Hill* was a milestone in UK hazardous waste control, this being the first time that a ship destined for demolition abroad had been prohibited from sailing. The procedure represented new ground for the Environment Agency (EA), who worked under tight time constraints to formulate procedure and then produce and serve the arrest documentation.[152] Thereafter, the case becomes one of an irregular, but increasingly common round of stated intentions, subsequently reversed once the ship had left UK jurisdiction, and changes of identity and ownership, whilst she proceeded to a breaker's yard.

Built originally as the liquefied natural gas (LNG) tanker *LNG Challenger*, the ship was to change its name three more times[153] before being sold by its then

[149] NGO Shipbreaking Platform (2013a).

[150] NGO Shipbreaking Platform (2013b).

[151] The French authorities were somewhat reluctant to keep the ship, envisaging abandonment by her owners and the cost to the state of demolishing the ship as per the *Sandrien* and (potentially) the *Sea Beirut*.

[152] Adding to the sensitivity of the issue was the case in 2009 of some 89 cases of supposedly scrap plastic waste exported to Brazil under the green list controls of the Basel Convention and without formal notice to the Environment Agency. When the Brazilian authorities discovered that the waste actually consisted of mixed municipal waste, the containers had to be returned at high cost to the UK at UK expense for appropriate disposal. Environment Agency (2009).

[153] March 1979 *LNG Challenger*, February 1987 *Pollenger*, December 1998 *Asake Maru*, September 2000 *Mystic Lady*, October 2007 *Hőegh Galleon*.

Japanese owners Mitsui OSK Lines to Hőegh LNG, initially as a source of spare parts for a sister ship, but subsequently subject to new repair methods, which allowed her to re-enter service as the *Hőegh Galleon*. In mid-2007, the ship was sold to Maverick LNG for conversion for an LNG liquefaction project, the vessel then acquiring the name of *Margaret Hill* and the start of a short but chequered final career. After the liquefaction project failed to occur, the ship lay idle in Spain for several months, before being laid up in Southampton in November 2008, registered under the flag of the Republic of the Marshall Islands. When the owners became bankrupt, title of the vessel passed to the mortgagee, Fortress Investment Group, in mid-2009.[154]

Late on 5 August 2010, information passed to the Maritime and Coastguard Agency (MCA) indicated that the vessel, which was considered likely to contain hazardous materials, was expected to leave Southampton for scrapping at some undefined destination in South Asia. An inspection by the MCA found hazardous materials including asbestos on board; these findings were reported to the EA.[155] The ship had value not only in the steel and other content that traditionally makes up a ship, but also in the 3,100 tonnes of nickel-steel[156] in its tanks and the estimated 520 tonnes of bunker fuel on board. Local sources also indicated that the vessel had been acquired for £5 million, but had a scrap value estimated to be around £17.2 million.[157] Waste ships sailing from England and Wales for scrapping overseas require the permission of the Environment Agency and from the equivalent regulators in the proposed country of destination; neither had been procured. Demolition of end-of-life ships containing hazardous waste may only be carried out in authorized yards either in the EU or in an OECD country. This was the first time that these powers had been implemented by the Agency, who therefore had no established procedure to follow.[158] After consultation with counsel at national level on the legal implications, rapid action by the Agency resulted in the production of a Stop Notice, issued under the provisions of the UK's *Transfrontier Shipment of Waste Regulations 2007*, Schedule 5, para. 5(2) (b) (ii), which enacted European hazardous waste legislation into UK law. This temporarily prohibited the vessel from leaving port until the EA had determined to its satisfaction that the ship was not intending to sail for demolition outside the UK and until the Agency had issued a written notice withdrawing the Stop Notice. In this respect, the powers of the EA to prevent ships from sailing is greater than that of the MCA, a simple belief rather than hard proof that illegal movements of hazardous waste were about to take place being considered sufficient.[159]

---

[154] Matthews (2010).

[155] Brown (2009), p. 8.

[156] Ecologist (2010).

[157] Hurrell (2009).

[158] Conversation with the Environment Agency staff involved in October 2009, just after the vessel had left Southampton.

[159] Comments from discussions with EA representatives, 3.12.2009.

Copies of the Stop Notice were served upon the ship's owners at Fortress Investment Group (UK), the ship's managers V Ships UK Ltd., and the ship's master on 7 August, just prior to the ship's intended departure. A copy was also issued to the Southampton port authority, Associated British Ports (ABP), which was thereby unable to provide a pilot to enable the ship to leave.[160] It is of note that the Stop Notice refers to the ship simply as 'waste,' and not 'containing waste.'

The authorised person issuing the Stop Note claimed:

> ...reasonable grounds to suspect that the vessel...is waste and that the TFS Regulations and/or Council Regulation (EC) 1013/2006...are not being, or are not likely to be, complied with.

The grounds cited for the suspicion were:

> ...the vessel was to be sold to a company and then shipped...from the United Kingdom with the intention of scrapping the vessel;
>
> ...the transport of the waste vessel...is either prohibited by the Community Regulation, or requires prior notification and written consent from all authorities concerned:
>
> ...the Environment Agency, as the competent authority of dispatch, has not given consent for any such transport:
>
> ...any transport of the waste to a place outside the United Kingdom in contravention of the prohibition...would be in contravention of the TFS Regulations.

After further investigations and negotiations with the owners and others, the EA issued a letter rescinding the Stop Order on 28 October 2009.[161] The new owners, US-based LNG specialist Waller Marine,[162] stated that their intentions were that the ship would be converted by DryDocks World in Dubai[163] to a floating treatment plant for liquid natural gas. By this stage, the expression 'going for repairs at Dubai' should set alarms sounding, following the cases of the *Blue Lady*, *Kaptain Boris* and doubtless numerous others. The ship finally left Southampton on 9 December, and UK jurisdiction shortly afterwards. It arrived at Jebel Ali in January and, after a short stay, was sold to a cash buyer, reputedly for transfer to India for demolition.

Despite calls from the NGO Shipbreaking Platform to the UK Government to establish a bond guaranteeing the authenticity of the re-use contract produced by the owners, the EA said that the law did not allow them to take a guaranteeing bond and, since the ship had left UK shores, it no longer fell within their jurisdiction. The

[160] Hurrell (2009).

[161] The ship was released after the Environment Agency claimed that it had received certificates and evidence of agreed deposits and a memorandum of agreement for the conversion. Barker (2010), p. 2.

[162] Waller Marine has alternatively been described not as owner, but as engineering consultant for the project. See Matthews (2010). Ownership was subsequently attributed to Waller Marine affiliate Polar Energy. Krishna (2010b).

[163] DryDocks World were to have been involved in the original plan to convert the ship when it was under Maverick ownership. With regard to this second plan, it announced that '*We have not entered any kind of contract and are not involved in any negotiations with the owner.*' Krishna (2010b).

movement from Dubai was now a matter for the authorities at Dubai.[164] It was hoped by the EA representatives that this case would serve as a message to UK ship exporters, since it would be difficult for the Agency to devote the same level of resources to other cases[165]; nevertheless, the model for similar action has now been established.

On 26 March 2010, Lloyd's List reported that it 'understands' that, although India was the initial destination (after departure from Dubai), the *Margaret Hill* will shortly head for China for demolition. Scrapping in China was regarded as perhaps offsetting to some extent the concerns associated with scrapping in India, since some of the Chinese yards are regarded as showing rather more environmental regard for their activities. Having been sold to Argo Systems for a reported US$10.2 million, the ship was renamed *Chill* and reflagged to the Comoros registry. By October 2010, the destination of the ship was still a matter of speculation, but an earlier entry on the August 2009 website of Marine and Commerce[166] lists the *Margaret Hill* under second hand ship sales as having been sold to Indian buyers. If a deception had been carried out on the EA similar to that involving the *Oceanic*/US EPA then, in the name of consistency, the Indian authorities should regard this (as well as the hazardous materials deemed to be on board) as appropriate grounds for denying entry to the ship, as they did with the *Platinum II*. The ship was finally demolished in Yinhu, Jiangmen Province, China.

5.11 Exxon Valdez/Oriental Nicety 2012

The case of the former *Exxon Valdez*, although involving in this instance just the one state of India, is included to reinforce the situation regarding the apparent lack of co-operation and co-ordination between Court and the various agencies and ministries involved. Despite the rulings of earlier cases, arguments and practices similar to those employed with the *Blue Lady* and *Platinum II*, scrapped earlier at Alang, continued to surface in subsequent cases.

The *Exxon Valdez* was completed in 1986 and was forever associated with the major oil spill in Prince William Sound, Alaska in 1989—indeed all of the articles which appeared about the proposed scrapping of the ship seemed to refer to the ship by her original name and the pollution with which she was associated, this adverse association proving a most useful handle for immediate recognition. Since the incident, the single-hulled tanker had been banned from entering in US and European ports, but had a varied career,[167] being converted to an ore carrier

[164] Krishna (2010b).

[165] Comments from discussion with EA representatives 3.12.2009.

[166] Marine and Commerce is a Turkish PR, publicity and promotions organisation.

[167] The ship subsequently became the *Exxon Mediterranean* (1990); *Sea River Mediterranean* (1993); *S/R Mediterranean* (1993); *Mediterranean* (2005) before conversion.

in 2008 and renamed *Dong Fang Ocean*. In 2011, the ship was finally sold to the cash buyers Global Marketing Services, buyers of the *Blue Lady* and *Platinum I*, and subsequently to M/s Best Oriental, a newly formed Hong Kong-based subsidiary of Priya Blue Industries Pvt. Ltd., the scrappers of the *Blue Lady*, for a reported US$16 million. Original intentions to scrap the ship in Singapore were later revised to Alang. Despite an initial ban from the Supreme Court against entering Indian territorial waters, the ship arrived in May 2012 under no flag, its former registration in Sierra Leone having expired.

Petitions ('Prayers') were presented to the Supreme Court by the environmental group Toxic Watch Alliance to prevent the demolition of the vessel and by the owners Best Oriental for its demolition.[168] Amidst various claims and counter-claims by the GCPB, the GMB, and the Ministries of Environment and of Shipping as to who had/should give permission for the ship to anchor at Alang for the purpose of inspection prior to demolition, the *Oriental Nicety* actually anchored off Alang for inspection in a scenario reminiscent of the *Blue Lady*, thereby presenting yet again a *fait accompli* to the Court.

After having been inspected by the GMB, the GPCB, Customs authorities and the AERB, an affidavit was filed in the Court claiming the ship to be free of any hazardous/toxic substances, although no IHM was presented to the Court. In a previous order given on 6 September 2007, the Supreme Court had recommended the production of a comprehensive Code to govern the procedure for allowing ships to enter Indian territorial waters and to beach at any Indian port for scrapping. However, such a Code had not been finalized and the GMB, the GPCB, and other nominated organizations had been empowered to oversee the arrangements for entry and inspection, thereby allowing them to give permission for the ship to be beached at Alang. In arriving at its judgment, the Court noted that the owner of the vessel was claiming very high demurrage costs while the ship was waiting for beaching. It therefore permitted beaching and dismantling to proceed, even though the requirements of the *Basel Convention* had not been complied with but subject to its former rulings on the entry of ships for scrapping,[169] stating that:

> ...once clearance has been given [*by the GPCB, GMB and AERB*] for the vessel to beach for the purpose of dismantling, it has to be presumed that the ship is free from all hazardous or toxic substances, except for such substances such as asbestos, thermocol or electronic equipment, which may be a part of the ship's superstructure and can be exposed only at the time of actual dismantling of the ship

[168] Supreme Court of India Civil Original Jurisdiction I.A. Nos. 61 and 62 of 2012 in *Writ petition (C) No. 65705 1995 Research Foundation for Science Technology and Natural Resource Policy v. Union of India and Ors.*

[169] The original rulings on ship demolition arising from the Writ Petition (Civil) No. 657 of 1995 filed by the Research Foundation for Science, Technology and Natural Resources Policy were made by the Supreme Court on 14 October 2003, 6 September 2007 and 11 September 2007.

However, it added that:

> . . .if any toxic wastes embedded in the ship structure are discovered during its dismantling, the concerned authorities shall take immediate steps for their disposal at the cost of the owner of the vessel, M/s Best Oasis Ltd., or its nominee or nominees

and concluded with the expectation that in future:

> the concerned authorities shall strictly comply with the norms laid down in the Basel Convention or any other subsequent provisions that may be adopted by the Central Government in aid of a clean and pollution free maritime environment, before permitting entry of any vessel suspected to be carrying toxic and hazardous material into Indian territorial waters.

It also directed the Government to ensure that the 1989 *Hazardous Wastes (Management & Handling) Rules* be brought in line with the provisions of the *Basel Convention* and with Articles 21, 47 and 48A of the Constitution.[170]

The Supreme Court's ruling was delivered on 30 July 2012 and on the 2 August 2012, the ship was beached at Alang. Pictures of the ship at anchor off Alang show yet another change of name, from *Oriental Nicety* to *Oriental N*.

5.12 Others

In addition to those cases examined above, there are numerous other, lesser known ships considered to be hazardous waste, and whose export for scrapping or whose arrival at intended shipyards have been denied. Although all ships carry some level of hazardous content, the ships that become embroiled in contention are those often targeted by the NGOs; Table 5.1 below lists just a few of them.

The reaction by environmentalists to the reflagging of two US-flagged ships on long term charter to the US Navy Military Sealift Command in August 2009 led to a revision of policy with regard to flag transfer of US vessels. The two ro–ro ships *PFC James Anderson* and *1st. Lt. Alex Bonnyman* were owned by the Wilmington Trust Co. Bank and were reported by the shipping press to be destined for Indian breakers after 5 years in commercial service, followed by 25 years in military use. Reflagged to the St. Kitts Nevis register and renamed *Anders* and *Bonny* respectively by their new owner Star Maritime Corp., the ships were allegedly to be employed in the Brazil–India sugar trade.[171] Reaction to this sale precipitated a move by MARAD in November which required all intended transfers to be cleared by the US EPA as to the fact that the vessel does not contain unacceptable levels of PCBs. This process also involves the outlay of up to US$150,000 in legal and consulting fees by the owner and a delay of 60–90 days for approval to be granted. This waiting time may be reduced if the owner voluntarily submits a statement

[170] Lucion Marine (2012).

[171] GreenDock (2009).

Table 5.1 Ships denied entry as hazardous waste

Ship	Type	Case	Date
Olwen	Naval tanker	Denied entry to Turkey Beached Alang 2001 as *Ken*	2000
Olna	Naval tanker	Denied entry to Turkey Beached Alang 2001 as *Kos*	2001
Silver Ray	Car ferry	Detained in Antwerp	2003
Novocherkassk[a]	Cargo ship	Denied entry to Turkey	2003
Hesperus	LPG tanker	Denied entry to Alang	2003
Alpha	Oil tanker	Denied entry to Bangladesh	2007
Asia Union	Cargo ship	Denied entry to Bangladesh	2011
Probo Koala/Gulf Jash/Hua Feng[b]	Tanker	Denied entry to Bangladesh. Reported en route to China	2011

Adapted from European Commission (2004)

[a]Purchased by the German company MSK, buyers of the *Sea Beirut* (see Sect. 5.2). Greenpeace (2010)

[b]As the *Probo Koala*, the ship was involved in the dumping of toxic wastes in the Gold Coast and was believed to be still carrying toxic waste residues

issued by an environmental consultant that the ship is PCB-free and the process may be by-passed altogether if scrapping takes place at a US facility, although the prices offered by US yards are considerably lower than in other markets.[172]

5.13 Summary Comment

These few selected case studies illustrate the way in which frequent changes of names and flags of ships intended for demolition, plus the frequent and often conflicting details of ownership, destination and intended use stated for some of these vessels, are employed to obscure the true identity and intended fate until a vessel actually arrives at its final destination for scrapping, possibly after more than one change of name and flag during its final voyage. Whilst it would be impossible for the *Clémenceau*, a state-owned vessel, to undergo such a routine of attempted disguise—a large aircraft carrier will always look like a large aircraft carrier—the cases of the *Riky*, the *Chill*, and the *Platinum II* are excellent illustrations of how shipowners seek to avoid liabilities for elderly and hazardous waste-filled ships.

On arrival at, for example, Alang, the sudden professed need to beach the ship rapidly for safety reasons appears to be a new, and sometimes successful, method of ensuring that the ship is rendered into a state where demolition remains the only feasible option—witness the *Blue Lady* and the attempted case of the *Platinum II*. Although procedures relating to pre-demolition inspections may be followed

[172] Joshi (2010). A request to MARAD for confirmation that this procedure is still in force elicited no response.

according to court directions, the questionable logic and selective interpretation behind some of the inspection reports as to what actually constitutes waste, and the repeated failures to observe the rulings of the Indian apex court, both by government Ministries and other agencies, all serve to ensure that once vessels have arrived at the beaches, their demolition is basically assured.

The *Platinum II* and the *Oriental Nicety* illustrated the way in which the activities of different government agencies could be set to operate in opposition to each other. This also appeared to be true in the case of the US Environmental Protection Agencies and MARAD, and in the cases of the various ministries of Bangladesh and of India. The manner in which various committees established by the Indian Supreme Court also illustrated how respective Ministries are able to implant their own representatives in a way that is less than impartial. With regard to the *Clémenceau* and the *Blue Lady*, this situation appears to have operated almost as a matter of course, governmental environmental and commercial agencies appearing to operate to their own individual interests rather than to a common doctrine. The pedestals upon which human and environmental rights are said by the Indian Supreme Court to sit—reference in particular the case of the *Blue Lady*—can at times appear to be of very modest height.

The efforts of an increasingly vociferous and growing band of campaigning NGOs fostered the growth of judgements from the national courts of Netherlands, Turkey, and France, as well as the UK's Environment Agency. Judgements and case law were based on European and Turkish waste laws, which in turn were based upon the provisions of *Basel*. These were a reflection of the growing international concern at the state of this industry with fragile regard for liabilities, yet the cases also demonstrate the difficulty of convincing some importing states of their obligations under the *Basel Convention* to reject such contaminated ships, even when direct appeals are made on a ministry-to-ministry basis, as in the cases of the *Riky*. Where it has proved impossible to hold the ship owner to account (as with the *Sandrien* and *Sea Beirut*), because of the ease with which beneficial ownership can easily and cheaply be obscured, ships have, as a consequence, ultimately been decontaminated or even disposed of by the intended exporting state, such disposals operations proving to be extremely expensive operations to the host state.

In an arena of argument and counter-argument, inter-government and intra-government disputes, selective use of evidence in court and hidden ownerships, this then was the situation that was extant at the introduction of the *Hong Kong Convention*.

References

Association of the Council of State and Supreme Administrative Jurisdiction of the European Union (2008) www.juradmin.eu/seminars/Brussels2008/France_en.pdf. Accessed 2 Oct 2010

Ban Asbestos Network of India (2007) Asbestos laden ship MV Al Arabia in Alang. http://banasbestosindia.blogspot.com/2007/11/asbestos-laden-ship-mv-al-arabia-in.html. Accessed 26 Sept 2010

Barker P (2010) Dubai bound Margaret Hill to leave Southampton Lloyds List, 30 October 2010, London

Basel Action Network (2002) Greenpeace 'most wanted' ship remains a thorn in port's side Lloyd's List, 11 February 2002, London

Basel Action Network (2005) Legal analysis of letter from Mr. Raja to Ms. Hedegaard regarding the legal application of the Basel Convention to the Kong Frederik IX (aka Ricky). www.ban.org/Library/BAN_analysis_Ricky.pdf. Accessed 3 Sept 2010

Basel Action Network (2006) Comments on statement of French Government on Clemenceau by the Basel Action Network on behalf of the greater coalition demanding return of Clemenceau to France for decontamination, BAN 24 January 2006. www.ban.org/Library/statement-of-frenchgovernment-on-clemenceau.pdf. Accessed 1 Oct 2010

Basel Action Network (2008a) Environmental victory for proper ship scrapping, BAN 16 May 2008. www.ban.org/ban_news/2008/080516_victory_for_proper_ship_scrapping,html. Accessed 1 Nov 2010

Basel Action Network (2008b) The Otapan principles, 12 May 2008. http://ban.org/library/o8o512_otapan_principles.html. Accessed 1 Nov 2010

Basel Action Network (2009) Toxic ship lands in India. BAN Media Release. www.ban.org/ban_news/2009/091022_toxic_us_ship_lands_in_india.html. Accessed 14 Apr 2010

Brown (2009) LNG Hill in scrap dispute. Fairplay, 20 August 2009, London

Central Technical Team (India) (2009) Report of the Central Technical Team constituted by the Ministry of Environment and Forests on the inspection of ship 'Platinum-II' anchored at Bhavnagar Anchorage Point, 23 October 2009

Culture Change (2012) U.S. ship disposal policy called 'shameful' following export of 'Exxon Valdez' and 'Delaware Trader' to Indian beaches. www.culturechange.org/cms/content/view/843/1/. Accessed 14 Oct 2013

Dutta M (2005) Environment Ministry plays ostrich. Meanwhile hazardous wastes are being dumped on Indian shores. Basel Action Network Toxic Trade News, 20 May 2005

Ecologist (2010) UK warned toxic ship would be scrapped on Indian beach, Ecologist, 18 February 2010. www.theecologist.org/News/news_round_up/418823/uk_warned_toxic_ship_would_be_scrapped_on_indian_beach.html. Accessed 30 Apr 2010

Environment Agency (2009) Chief Executive's update on key topics. open board meeting date, 15 September 2009. www.environment-agencyuk/static/documents/Leisure/CEUpdateSeptember v2.doc+defra+tanker+margaret+hill. Accessed 20 Dec 2012

European Commission (2004) Oil tanker phase out and the ship scrapping industry. EC Directorate General Energy and Transport P-59106-07 Final 2004

Ghatwai M (2005) Nod after committee inspected facilities at ship-breakers plot. www.ban.org/ban_news/050517_nod_after.html. Accessed 25 Aug 2005

GreenDock (2009) Inconsistency in the United States. www.greendock.nl/index.php?menu=actueel&page=actueel58. Accessed 14 Apr 2010

Greenpeace (2002a) Shipbreaking. News. Malpractice at ship-for-scrap. Sandrien – Crew and environment victim of ship dealer. www.greenpeaceweb.org/shipbreak/news14.asp. Accessed 1 Oct 2010

Greenpeace (2002b) Greenpeace takes French government to court for sending toxic ship to Turkey. www.greenpeaceweb.org/shipbreak/news37.asp. Accessed 1 Oct 2010

Greenpeace (2002c) Shipbreaking – the Sea Beirut case. A case of illegal traffic in hazardous waste. www.greenpeace.org/shipbreak/Factsheet_Sea_Beirut_Remarks. Accessed 12 Jan 2006

Greenpeace (2005a) Toxic ship Riky to be mercilessly driven out. www.greenpeaceweb.org/shipbreak/news107.asp. Accessed 1 Oct 1010

Greenpeace (2005b) Asbestos for India: France considers aircraft carrier's fate. www.greenpeaceweb.org/international/news/ghost-ship-121205. Accessed 1 Oct 2010

Greenpeace (2006) May we have the truth please? Greenpeace, 5 January 2006. www.greenpeaceweb.org/india/may-we-have-the-truth-please. Accessed 28 Feb 2011

Greenpeace (2010) Turkish authorities turn back European ship attempting to illegally dump waste in Turkey on Greenpeace warning. www.greenpeace.org/shipbreak/news54.as. Accessed 1 Oct 2010

Hurrell S (2009) Global investment group in toxic ship row. www.clickgreen.org.ik/news/international-news/12530-global-invetsment-group-in-toxic-ship-row.htm. Accessed 16 Oct 2009

IMOWatch (2009) Fraudulent certificate of Platinum-II. http://imowatch.blogspot.com/2009_11_01_archive.html. Accessed 8 Nov 2009

IMOWatch (2010) Platinum II case may change Indian vessel scrapping rules, 16 March 2010. http://imowatch.blogspot.com/2010/03/platinum-II-case-may-change-Indian.html. Accessed 11 June 2010

India Together (2006) The scrapping of Riky, 23 March 2006. www.indiatogether.org/2006/mar/env-riky.htm#cont. Accessed 1 Sept 2010

Indian Platform on Ship Breaking (2009) Letter from IPOS to the US Government via the US Consulate General, Mumbai and the US Embassy, New Delhi, Dated 10 October 2009. https://imowatch.blogspot.com/2009/10/letter-to-us-govt-on-convicted-toxic.html. Accessed 3 Jan 2010

International Centre for Trade and Sustainable Development (2006) Chirac recalls asbestos-laden ship from India, Orders inquiry, 22 February 2006, Geneva

Joshi R (2010) Scrapping US-flag vessels can cost owners US$150,000 in fees. Lloyd's List, 4 October 2010, London

Krishna G (2006) French Apex Court rules, Clemenceau recalled. India Together, 16 February 2006. www.indiatogether.org/. Accessed 10 Oct 2010

Krishna G (2007) Setting a precedent for trafficking hazardous waste. India Together. www.indiatogether.org/cgi-bin/tools/pfriend.cgi. Accessed 2 Nov 2010

Krishna G (2010a) Gujarat's report on channel blockage due to US toxic ship is bogus. www.mynews.in/Gujarat's_Report_on_channel_blockage_dueto_US_toxic. Accessed 14 Apr 2010

Krishna G (2010b) Possible arrival of UK's dead and toxic vessel's arrival in India. www.mynews.in/News/Possiblearrival of UK's dead and toxic vessel's arrival. Accessed 2 Mar 2010

Krogstrup E, Arleth KKN (2006) Ship scrapping – a floating scenario. Final Thesis, University of Roskilde, Roskilde

Kumar S (2004) Killer Clemenceau: the ecocrime that almost was. Financial Express, 29 August 2004. www.financialexpress.com/fe_full_story.php?content_id=67239. Accessed 30 Oct 2010

Lloyd's List (2008) Able UK wins contract to dismantle Clemenceau. Lloyd's List, 2 July 2008, London

London C (2006) The Clemenceau: a wasted end-of-life? Environmental liabilities 14(2):64

Lucion Marine (2012) India's Supreme Court bans entry of foreign ships with toxic waste, 10 July 2012. www.greenpasport.net/india%e2%e80%e99-supreme-court-bans-foreign-ships-with-toxic-waste/

Matthews S (2010) Speculation mounts over Margaret Hill conversion plan. All Business. www.allbusiness.com/energy-utilities/oil-gas-industry-oil-processing-products/13932347-1.html. Accessed 2 Mar 2010

Mediavigil (2007) Waste follows the path of least resistance, 4 November 2007. http://mediavigil.blogspot.com/2007_11_01_archive.html. Accessed 26 Sept 2010

Mendez R (2007) Environment detains a boat in Almeira to prevent scrapping in India. El Pais, 29 September 2007, Madrid

Mines and Communities (2007) Blue Lady owner misled the Court & Ministry. Government admits radioactive material on Blue Lady. www.minesandcommunities.org/article.php?a=315. Accessed 14 Apr 2010

Ministry of Environment and Forests (India) (2009) Full report on Platinum II. www.Moef.nic.in/downloads/public-information/Final_Platinum-II_Report.pdf. Accessed 14 Apr 2010

NGO Shipbreaking Platform (2006a) Star Cruises Ltd. and Norwegian Cruise Lines: deceiving Germany and violating international law in the export of the SS Norway to India. NGO

Shipbreaking Platform, 30 June 2006. http://ban.org/library/Star_Cruises_Deception_Report_Final.pdf. Accessed 14 Apr 2010

NGO Shipbreaking Platform (2006b) Evidence of deception by owners of Blue Lady emerge. Merinews. www.merinews.com/catFull.jsp?articleID=211&category=Nation&catID=2. Accessed 2 Oct 2010

NGO Shipbreaking Platform (2006c) Govt. bans toxic ship's entering Bangladesh. Toxic Trade News, 17 February 2006. www.ban.org/ban_news/2006/060217_bans_ship.html. Accessed 4 May 2006

NGO Shipbreaking Platform (2006d) Comments on the Indian committee inspection report on the hazardous materials on board the SS Blue Lady, 31 July 2006. BAN 2006. www.ban.org/NGO_Platform_Critique_on_TC_Inspection_Report_Final.pdf. Accessed 1 Mar 2011

NGO Shipbreaking Platform (2013a) Finland permits illegal export of hazardous waste. www.shipbreaking.platform.org/finland-permits-illegal-export-of-hazardous-waste/. Accessed 30 Sept 2103

NGO Shipbreaking Platform (2013b) Letter to Finnish Minister of the Environment. www.shipbreakingplatform.org/shipbrea_wp2011/wp-content/uploads/2011/11/Letter-FINLAND-MAY-101.pdf. Accessed 30 Sept 2013

Norohna F (2006) Toxic warship not allowed to disrupt France-India Summit EN-Newswire, 26 February 2006. www.en-newswire.com/feb2006/2006-02-20-04.asp. Accessed 10 Oct 2010

Orellana M (2006) Shipbreaking and the Clemenceau row. ASIL Insight (Am Soc Int Law) 10(4)

Pandi V (2009) Ship graveyard mourns loss of business. National Law School of India University. www.nlsenlaw.org/copy_of_news/ship-graveyard-mourns-loss-of-business. Accessed 14 Apr 2010

Ramachandran S (2009) Toxic alert as US ship heads for India Asian Times. www.atimes.com/atimes/South_Asia/KJ24Df01.html. Accessed 18 Jan 2010

Sethi N (2009) It's official: Platinum II ship is toxic. Times of India, 6 November 2009, Delhi

Specis K, Tan HE (2008) SS Norway ex SS France. SS Maritime. www.ssmaritime/Norway-Timelines.htm. Accessed 14 Apr 2010

SSMaritime (2009) SS Independence. www.www.ssmaritime.com/ss-independence-constitution.htm. Accessed 26 Sept 2010

Stanford PH (2007) Keynote address remarks delivered at the Stanford Law Review Symposium in global constitutionalism at the Stanford Law Centre, 16 February 2007. Stanford Law Review (2007) 59 1155

Vlierodam Wire Ropes Ltd. (2004) Dutch carry chem. tanker scrap costs. Daily Shipping Newsletter 2004 – 237. www.biblio.org/maritime/Pdf/scheepvaartnieuws/2004/novem/237-16-11-2004.PDF Accessed 1 Oct 2010

Zarach S (2006) The Clemenceau. www.bimco.org/Print.aspx?itemId=[B4B63544-A707-4255-9243-7AE5A12AF. Accessed 15 Oct 2008

Chapter 6
The Hong Kong Convention 2009

In 2009, the *International Convention for the Safe and Environmentally Sound Recycling of Ships* (the *Hong Kong Convention*) was approved by 63 Members of the IMO; this chapter will examine the development and content of the Convention.

6.1 Background to the Convention

Attempts to bring order to shipbreaking began when the *Basel Convention* entered into force in 1992 and its provisions were deemed relevant (at least by the proponents of *Basel*) to the industry. However, the lack of specific provisions for end-of-life ships, particularly with regard to identifying the point at which a ship might be regarded as waste, led to an ongoing debate as to its applicability. Although the provisions of the Convention, and its subsequent amendment, were incorporated into European legislation via the *Waste Shipment Regulation*, in 2011 the EC announced that the problems of enforcing this measure effectively made it inoperable with regard to end-of-life shipping. Guidelines for the industry were produced by the Basel Convention Secretariat in 2002 in order to inform the shipbreaking states on the promotion of the measures for environmentally sound management of hazardous waste at their various facilities. Further guidelines were issued in 2004 by the ILO,[1] which had a direct concern with the occupational health and safety of the workers in the industry, but both sets of guidelines were established on a non-mandatory basis.

The subject of a mandatory convention dedicated to regulating the activities of the industry was first brought before IMO at the 42nd session of its Marine Environmental Protection Committee (MEPC42) in November 1998 and the importance of the IMO's role in this regard was agreed at a following meeting of

[1] These Guidelines were produced as a supplement to the 1981 ILO Convention on Occupational Health and Safety and issued at the ILO's 289th session in March 2004.

M. Galley, *Shipbreaking: Hazards and Liabilities*, DOI 10.1007/978-3-319-04699-0_6,

MEPC45. By March 2002, at MEPC47, agreement had been reached that the IMO should compile further recommendatory guidelines that would also acknowledge the work produced by the other organisations. These guidelines—*IMO Guidelines on Ship Recycling*—had been finalized by MEPC49 (July 2003) and were adopted at the 23rd assembly of the IMO that year.[2] They were directed at a wide range of stakeholders in the shipping industry, including flag, port and recycling states, ship owners, ship builders and equipment suppliers. It is considered that the need for the *HKC* may not have arisen had the provisions of the *IMO Guidelines* actually been followed—the ongoing conditions at the breakers was brought to the attention of the *Basel Convention* by the work of the NGOs.[3]

By the 53rd session of the MEPC (July 2005), the need for further mandatory and legally-binding global measures was agreed and at the 24th assembly, the IMO called for a new legally-binding instrument on ship recycling to be developed by the MEPC by 2008–2009.[4] A draft text had been submitted by Norway at the MEPC54 (March 2006), and a Joint Working Group on Ship Scrapping was formed to work on the development of this draft, with a Correspondence Group continuing the work between sessions of the Working Group. Attempts to remove the problems perceived on the differences existing between the three set of guidelines and the gaps in the regulatory framework were made in the new instrument. An acknowledgement of the contribution of the BC and ILO to the development of the new instrument was included in the preamble to the Convention and it has been suggested that this integrated working between institutions is likely to remain the manner in which regulation of the industry is likely to continue.[5] For the first time, an IMO convention was directed at mandatory standards for land-based activities. To this point and in the ensuing MEPC meetings, the shipbreaking industry appeared to play a minor role in the development of the Convention and its guidelines.

Discussions by the author at the IMO prompted the comment that an attempt to introduce bilateral agreements in a fashion to those incorporated into the *Basel Convention* was made by the USA when the *HKC* was being formulated; this move was firmly defeated by other nations who did not wish to see the introduction of measures beyond the basic provisions of the Convention.

At its 55th session, the MEPC agreed to request the IMO Council to schedule a five day international conference in 2008–2009 to consider the new Convention, and this duly took place at Hong Kong in May 2009, when the Convention was formally adopted. The gestation period of 3 years and 2 months from first draft of the text to the formal adoption of the Convention was considered something of a record for the IMO.

---

[2] Adopted under Resolution A.962(23), later amended at the 24th assembly in December 2005 under Resolution A.980(24).

[3] Comment arising from discussion with the government adviser 17.1.2011.

[4] Resolution A.981(24).

[5] Harrison (2009), p. 736.

6.2 Structure of the Hong Kong Convention

It is not intended to define or describe here the entire contents of the Convention on an article-by-article basis, but this section will examine its main provisions.

After the Preamble, the *HKC* consists of:

21 Articles that define the main legal mechanisms, including procedures for amending the Convention (Article 18)
25 Regulations covering technical requirements, consisting of 4 Chapters:

i. General (Regulations 1–3)
ii. Requirements for ships (Regulations 4–14)
iii. Requirements for ship recycling facilities (Regulations 15–23)
iv. Reporting requirements (Regulations 24–25)

7 appendices covering lists of hazardous materials and required formats for defined certificates and other documents
6 non-mandatory sets of guidelines to support the Convention

although how guidelines open to national interpretation equate to 'unified' interpretation or how voluntary guidelines give rise to a 'legally binding instrument' are perhaps questions of semantics given the subjectivity of interpretation that has occurred elsewhere.

Amongst the various definitions, the definition of 'shipowner' includes:

> those who have ownership of the ship **for a limited period** pending its sale or handing over to a Ship Recycling Facility.[6] (emphasis added)

This definition essentially covers the role of the cash buyer, who may or may not decide to register his vessel for its final voyage to the breakers, the decision not to register effectively removing the role of a flag state for that particular ship.

6.2.1 *General Content*

Article 1 on General Obligations requires that all parties follow the provisions of the Convention in order to:

> Prevent, reduce, minimize and, to the extent practicable, eliminate accidents, injuries and other adverse effects on human health and the environment caused by Ship Recycling...

> ...to endeavour to co-operate for the purpose of effective implementation of, compliance with and **enforcement of** this Convention... (emphasis added)

and to

[6] Regulation 1 (8), 'shipowner' not being included in the definitions listed within Article 2.

. . .encourage the continued development of technologies and practices which contribute to safe and environmentally sound Ship Recycling. . .

This Article is significant in that it recognises the damage that emanates from the current practices and procedures of the industry and that the aim is to produce a regime that is not static, but undergoes continual improvement. The reference to enforcement is of interest in that there are no defined enforcement penalties for infringement of the standards of operations that are established and policed by the individual states themselves (the IMO having no enforcement capabilities), the guidelines upon which these may be based themselves being non-mandatory but open to national interpretation. Administrations are to enact appropriate legislation for the regulation of ships and recycling facilities under their jurisdiction and are required to take proceedings against reported miscreants, but this also remains a matter under national sovereignty (although other states may take measures against violating ships such as prohibiting them from entering their ports).

Article 3 defines the Convention as being applicable to '*ships entitled to fly the flag of a Party or operating under its authority*.' Excluded from such ships are (as appearing in other IMO conventions) warships, naval auxiliaries, or other ships owned or operated by the government for the time being on non-commercial service. Also excluded from the scope are ships of less than 500GT or ships operating throughout their life in waters that are under the sovereignty or jurisdiction of a Party.[7] Further references to this Article will be given below. The Article also includes Ship Recycling Facilities operating under the jurisdiction of a Party within its scope, the Parties having responsibility for ensuring compliance with the Convention of such facilities within their jurisdictions (Article 4).

The new Convention leans on the provisions of the earlier IMO guidelines and a report to the MEPC's 49th session[8] contained three concepts which were entered into the draft Convention, namely:

- the concept of an inventory of hazardous materials (IHM)—often referred to previously as the 'green passport'
- licensing and regulation of the recycling yards, and
- the responsibility of shipowners (although this did not extend to any provisions for pre-cleaning before despatch to the breakers).

The report also referred to four key principles which should be observed in the new instrument (although not all were subsequently incorporated):

- the minimisation of hazardous materials and waste
- recyclability of materials should not be the sole criteria for their acceptability

---

[7] The Convention adds that, although the latter grouping of ships may not be subject to its provisions, Parties should adopt measures that require such ships to operate in a manner consistent with the Convention. On the other hand, ships which fly the flags of non-Parties should not be the recipients of favourable treatment.

[8] Annex 1 to the report of the co-ordinator of the Correspondence Group to the MEPC, 28 March 2003.

- the need for giving priority to safety and environmental protection during a ship's operational life, and
- the 'polluter pays' principle.

Much of the Convention consists of details of a series of inspections, verifications and documentation defined for the numerous stages between the commissioning of a new vessel (with regard to its IHM); the owner's notification of a ship ready for demolition; approval of a breaking plan; and the breaker's certificate of completion of the demolition. Article 13 defines the requirements for support via the transfer of management systems and technology relating to ship recycling between the Parties; a similar measure is contained within *Basel*.

6.2.2 Provisions for Ships

The Convention basically seeks to ameliorate the risks currently endemic within the industry firstly by defining technical standards for ships prior to breaking and secondly by setting a regulatory framework for the recycling facilities; it says little about the actual decontamination process. With regard to ships, Part A, Regulations 4–7 of Chapter 2 of the Convention, covers the design, construction, operation and maintenance of ships, and then addresses the preparation of ships for recycling, surveys and certification. Part B, Regulations 8 and 9, covers the preparations for ship recycling and Part C, Regulations 10–14, relates to surveys and certificates.

Many of the provisions relating to the hazardous contents of ships were included in the earlier *IMO Guidelines*; in the Convention they relate to the identification of hazardous materials, their prohibition/restriction in the construction and repair of ships (Regulation 4) and their documenting in an IHM (Regulation 5) to be kept on board the ship—the 'Green Passport' of the Guidelines.[9] Details of the locations and approximate quantities of these materials should also be included. The list of prohibited/restricted materials for new builds is contained within Appendix 1 to the Convention—Controls of Hazardous Materials. Appendix 2 consists of a Minimum List of Items for the Inventory of Hazardous Materials. The IHM itself consists of three parts; Part I is a listing of hazardous materials, as detailed in Appendix 1 and 2 of the Convention, which are on board or incorporated into the vessel's structure.

Parts II and III relate to operationally generated wastes and to stores on board respectively; these two parts are to be completed when the ship is ready for disposal. Since the levels of stores and wastes detailed are those remaining on board ship just prior to its departure to the breakers, these may well be changed by the time the vessel finally arrives at its destination. The question was therefore put in one of the interviews as to whether that could prompt a close re-appraisal of the

[9] Although the use of asbestos has long been a substance prohibited in ship construction, its presence is still found in both existing and new ship constructions—see the case of the *Caroline Essberger* 2009 at Sect. 3.2.9.

contract when the vessel finally arrived if, in the interim, trade conditions had actually deteriorated—not an uncommon practice for shipbreakers when the level of their return seemed to be suddenly and adversely threatened. The response was that this was not impossible, but the levels noted were regarded (at least by the sellers) as just approximations rather than precise measurements.[10] It remains to be seen whether such an interpretation has universal acceptance with shipbreakers in times of economic stress.

There are distinct practical liabilities in compiling a fully comprehensive inventory—for example listing all instances of the use of lead in solder; in such circumstances, the phrase 'as far as practical' is likely to be employed.[11] It should be noted that IHMs are not only of relevance at the point of demolition, but may apply equally during the operation, repair and maintenance of a ship, advising crew and shipyard workers of the presence of hazardous materials.

The optimum time for compiling an IHM is obviously when the vessel is under construction. Data is to be provided by the shipbuilders, who themselves are reliant upon data supplied by a wide range of equipment and materials suppliers. The difficulties involved with compiling data relating to substances incorporated into the materials and equipment which may come from a vast range of suppliers at all levels, not simply Tier 1 suppliers,[12] and in a common, defined format, should not be underestimated.[13] Whilst it may just be possible to eliminate all traces of prohibited substances which may be introduced during the manufacturing processes, revisions to design and construction practices may be required.[14] Once compiled, this inventory is to be maintained and updated in accordance with guidelines, as repairs and modifications to the structure of the ship over its lifetime change the overall makeup of its content with regard to hazardous materials. Errors in the IHM then become the responsibility of the owner.

There also lies the possibility that shipowners placing orders for new vessels may require that those vessels be compliant with the requirements of the Convention, even though the Convention may not yet be in force—particularly since the optimum time for compiling an IHM is during the construction of a ship. In its analysis of possible scenarios relating to the introduction of the Convention, the Superyacht Builders Association (SYBAss) offers various alternative strategies for

[10] Discussion with recycling facilitator, 2.5.2012.

[11] Townsend (2012).

[12] Tier 1 suppliers are those that supply directly to the production process rather than suppliers further down the supply chain.

[13] To this end, a Compliance Data Exchange Scheme is being researched and compiled in order to facilitate and standardise a database format that will accommodate product information that has to be passed up from suppliers to customers. Given the mass of data that will be involved for the production of any vessel, a high degree of automation will be required but it is not deemed likely that 100 % coverage can be provided. Schneider (2011).

[14] SYBAss (2011). If the compilation of data required during the construction of even a large motor yacht is considered problematic, then the degree of difficulty that one might expect in constructing a large ocean-going ship must be at a level many times greater.

its members, including vertical integration, obtaining information directly from its Tier 1 suppliers (with the attendant problems of some suppliers being unable to provide full information or unwilling so to do because of possible disproportionate costs involved) or to operate via a third party service provider, who may also experience the same problems. A third option may be for the industry to act as one in the compilation of data, difficulties here also adding problems of delay in establishing a world-wide system.

The procedure for existing ships is somewhat more complicated in that the inventory is more difficult to compile on an *ex post* basis. Owners will have some 5 years to establish IHMs for their vessels from the time the Convention enters into force, or by the time a ship is to be despatched to the breakers, if this is earlier. Visual or sampling checks are to be undertaken in order to develop an IHM and the results entered to a plan of the vessel. Under the Convention (and the proposed Regulation—see Chap. 7) a handful of cash buyers will have all the liabilities of shipowners (albeit for a limited time) except trading and hence will ultimately be responsible for ensuring the provision of IHMs for the breakers.[15]

Once an IHM has been compiled (both for new and existing ships), an initial survey is required before an International Certificate on Inventory of Hazardous Materials (ICIHM) is issued by the relevant Administration, with subsequent renewal surveys required to ensure that the IHM still meets the requirements of the Convention, these renewal surveys to be carried out at intervals specified by the Competent Authority (CA) but not exceeding 5 years.[16] The verification of a valid ICIHM may be made by the appropriate officers of Parties on ships subject to the Convention which enter their ports.[17] Whatever the final date for the entry into force of the Convention, there is likely to be an enormous demand for IHM certificates and this in turn may present a bottleneck if sufficient trained and approved hazardous materials experts are not available.[18] Despite the difficulties of preparing such a listing for existing ships, DNV had by 2009 already compiled some 50 IHMs for operational ships, prompted by the earlier requirements of the *IMO Guidelines* 2003.[19]

Once a ship is deemed ready for demolition, the owners must select an authorized Ship Recycling Facility (SRF) within a Party state that is capable and

---

[15] Townsend (2012).

[16] Further general or partial surveys may be carried out at the request of the shipowners after changes or replacement to the ship's structure or fittings to ensure that the ship continues to comply with the Convention and that Part 1 of the IHM is amended as necessary—Regulation 10.1.3. However since certifiers have their own inventory software, each classification society tends not to accept the certificates of others.

[17] These inspections may also include verification of a valid International Ready for Recycling Certificate for ships *en route* to the breakers.

[18] In 2009, the publication *Tanker Operator* estimated that should the Convention be in effect by 2013, an estimated 50,000 ships worldwide would require a valid IHM certificate. Tanker Operator (2009).

[19] Martinsen (2009), p. 12.

authorized to deal with the hazardous materials as are listed in the ship's IHM, and provide copies of the IHM, ICIHM and any other relevant information to that facility. At this stage, Parts II and III of the IHM are to be completed and the flag state Administration notified of the intention to recycle the ship. A formal, ship-specific Ship Recycling Plan (SRP)[20] is to be completed by the facility and is to include details of the facility's provisions for Safe-for-entry and Safe-for-hot-work activities, together with details of how the types and quantities of hazardous materials identified in the IHM are to be managed.[21] If the facility is authorized to handle the hazardous materials detailed in the IHM, then prior removal of those materials shall not be required unless specified by the facility in the SRP.

This plan is to be approved, either on an explicit or tacit basis,[22] by the relevant Competent Authority[23] and its receipt acknowledged to the facility, the ship owner and Administration. When an approved SRP has been received by the owner, a final survey is to be arranged by the owner before the ship is taken out of service in order to verify that the IHM is correct and reflected appropriately in the SRP. Successful completion of that survey enables the flag state to issue to the owner an International Ready for Recycling Certificate (IRRC).

6.2.3 Provisions for Ship Recycling Facilities (SRFs)

Chapter 3 of the Convention defines the responsibilities of the Parties towards SRFs in their jurisdictions. Regulation 15 requires that states are responsible for establishing the appropriate legislation, regulations and standards to ensure that the facilities are '*designed, constructed and operate in a safe and environmentally sound manner*.' They are further required to establish measures for the effective inspection, monitoring and enforcement of the facilities; such measures may include powers of entry and sampling and audits by the Competent Authority

[20] A number of Indian yards are already submitting breaking plans to owners. Comment from discussions with shipbreakers' representative, 24.2.2010.

[21] The requirements for the details of a SRP are contained in Regulation 9 of the Annex to the Convention. The two parts of the guidance consist of: general guidance on information that should be gathered and reviewed by the SRF in order to develop the plan (section 3: General); and guidance for the recommended content of a ship-specific SRP (section 4: Framework of SRP).

[22] '*Where a Party requires explicit approval of a SRP, the CA shall send written notification of its decision to the SRF, shipowner and Administration. Where tacit approval is required, acknowledgement of receipt of the SRP shall specify the end date of a 14-day review period. Any written objection to the SRP shall be sent to the SRF, shipowner and Administration within this period. If no such written objection is notified, the SRP shall be deemed to be approved.*' Regulation 10.4.1 and 10.1.2.

[23] 'Competent Authority(ies)' is defined as '*a governmental authority or authorities designated by a Party as responsible, within specified geographical area(s) or area(s) of expertise, for duties related to Ship Recycling Facilities operating within the jurisdiction of that Party as specified in this Convention.*' Article 2.3 Definitions.

(CA), the results of any audits being passed to the IMO. All these provisions should be in accordance with the guidelines contained within the Convention, but in practice will be subject to the state's own interpretation.

The authorization of an SRF is to be carried out by the state's CA on the basis of a site inspection and verification of prescribed documentation (Regulation 16), the format of an approved Document of Authorisation to Conduct Ship Recycling (DASR) being defined in Appendix 5 of the Convention. Refusal of an inspection on the part of the SRF is to result in the suspension or withdrawal of the authorization, which is otherwise valid for a period as defined by the CA, but not to exceed 5 years. Any incidents or actions at the SRF which contravene the conditions of the authorization are to be notified to the CA and may also result in suspension or withdrawal of the authorization or require corrective action.

The facility itself is to establish such management systems and procedures as to eliminate or minimize to the extent practicable the adverse environmental effects of its operations and to remove health risks to the workers *and* to the population in the vicinity (Regulation 17). Such authorized facilities shall only accept ships which are in compliance with the Convention and that they are authorized to recycle. They shall also have the appropriate authorization documentation available for any ship owner considering using that facility. The SRF Board or governing body is required to prepare a Ship Recycling Facility Plan (SRFP) (as distinct from a Ship Recycling Plan for a specific vessel), which details the measures adopted to ensure a number of specified criteria, including (but not limited to) protecting the health and safety of the workers and of the environment; the roles and responsibilities of employers and workers; the provision of information and training of workers; a system of monitoring of performance and record keeping; a system for reporting discharges, emissions, accidents and injuries etc.; and a plan for emergency preparedness and response (Regulation 18). These items are further defined in Articles 19–24 of the Convention.

Once a ship has acquired an International Ready for Recycling Certificate (IRRC), the facility may not start recycling until it has reported the planned start of the recycling to its CA and included a copy of the IRRC. Once demolition has been completed, the facility is to issue a Statement of Completion (SC) to its CA, who will in turn copy the SC to the Administration issuing the IRRC for the ship.

A summary chart of the basic flow of information and authorizations required prior to a ship being despatched for demolition is shown below in Fig. 6.1

6.2.4 Non-Party Ships and Non-Party Facilities

It is the intention of the Convention to regulate

- the design and construction of ships flying the flags of the Parties with regard to the use of potentially hazardous materials used in the construction and repair of

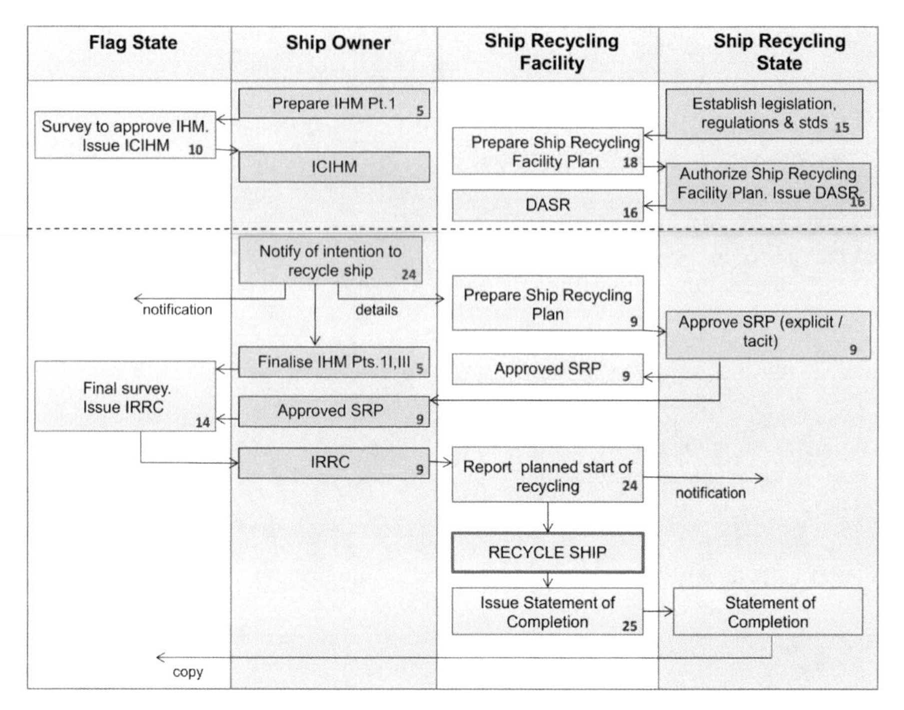

DASR Document of Authorization to Conduct Ship Recycling, *IHM* Inventory of Hazardous Materials, *ICIHM* International Certificate of Inventory of Hazardous Materials, *SRP* Ship Recycling Plan, *IRRC* International Ready for Recycling Certificate, 5 Convention article. Source: Galley (2013)

Fig. 6.1 *Hong Kong Convention* flowchart

the ships and the production and maintenance of a verified inventory of such materials, and

– the inspection and authorization of ship recycling facilities within the jurisdiction of the Parties with regard to their procedures for the handling, storing and disposal of such hazardous materials.

However, after the Convention has entered into force, some ships demolished in Party facilities may be non-Party ships. Whilst authorized SRFs may only accept ships that comply with the Convention (Party ships) as per Regulation 16.1, or ships that meet with the requirements of the Convention (non-Party ships) as per 17.2, Mikelis[24] offers that the cost for a non-Party ship to actually meet the requirements of the Convention is not more than US$3 per LDT (US$30,000 for a 10,000 LDT Panamax vessel, taking this as an example). It is, however, common practice for ships to be sold, de-registered and reflagged, Party ships thereby becoming non-Party ships which may then be demolished at a non-Party SRF. The cost of changing flag is estimated at just US$1 per LDT., or around US$10,000 for a 10,000 LDT Panamax.[25]

[24] Mikelis (2010a).

[25] Mikelis (2010a).

Mikelis concluded that under the *HKC*:

- Party facilities will be able to accept both Party ships and non-Party ships (the latter at a compliance cost to the owner of around US$30,000)
- Non-Party facilities will be able to accept Party ships (at a re-flagging cost of around US$10,000) and non-Party ships.[26]

In this respect, the *HKC* does not act as much of a barrier to those wishing to avoid the requirements of the Convention.

6.2.5 Guidelines

Whilst the Convention itself was adopted in 2009, the regulations of the Convention are based upon a series of six sets of guidelines, intended to provide for uniform interpretation and procedures, but their non-mandatory status still leaves their interpretation, implementation and enforcement a matter for each individual Party.

The subjects covered include:

- Guidelines for the development of an Inventory of Hazardous Materials (the Inventory Guidelines)
- Guidelines for safe and environmentally sound ship recycling (the Facility Guidelines)
- Guidelines for the development of the Ship Recycling Plan (the SRP Guidelines)
- Guidelines for the authorization of Ship Recycling Facilities (the Authorization Guidelines)
- Guidelines for surveys and certification
- Guidelines for inspection of ships.

Figure 6.2 below shows the progress of adoption of the various guidelines.

6.2.6 Enforcement and Dispute Settlement

The IMO has no powers of enforcement or sanctions with regard to non-compliance with its treaties; such provisions within the *HKC* are matters which are basically left to individual Members. Measures to ensure that the operation of ships flying Member States' flags and the authorization of SRFs within their jurisdictions are to be enacted into national legislation (Articles 4–6), which shall also prohibit violations of the treaty (Article 10). Detection and enforcement of violations is based upon co-operation between the Parties (Article 9), although sanctions, which

[26] This is assuming that the penalties included in the proposed European ad hoc Regulation (see below) for such actions are not in force.

MEPC Session:	MEPC 59 Jul 2009	MEPC 60 Mar 2010	MEPC 61 Sep-Oct 2010	MEPC 62 Jul 2011	MEPC 63 Mar 2012	MEPC 64 Oct 2012
Guidelines for the devel. of Invent. of Hazmats (Inventory Guidelines)	Adopted			Amended		
Guidelines for safe and env. sound ship recycling (Facility Guidelines)					Adopted	
Guidelines for the devel of the Ship Recycling Plan (SRP Guidelines)				Adopted		
Guidelines for the authorization of Ship Recycling Facilities (Authorization Guidelines)					Adopted	
Guidelines for survey and certification						Adopted
Guidelines for inspection of ships						Adopted

Fig. 6.2 The adoption of *Hong Kong Convention* guidelines. Source: Adapted from Mikelis (2011)

are to be of sufficient severity as to discourage such violations, are not defined with regard to their nature and extent. All these issues are matters for subjective interpretation by Members[27]—compare this with the provisions of the proposed ad-hoc European Regulation, which sets out quite specific penalties for owners of ships seeking to evade the use of approved SRFs (see Chap. 7). Neither does the Convention contain any defined process for the settlement of any disputes arising between States, other than '*by negotiation or any other peaceful means agreed upon by them,*' including judicial settlement (Article 14).

However, the *HKC* is characterised by an aspect that puts it somewhat apart from other IMO conventions. The new Convention introduces a comprehensive system of surveys, planning, approvals and confirmations, which in turn imposes direct involvement and assent by the exporting marine administration where previously this did not exist. The performance of individual shipbreakers will therefore have to be acceptable to a wider international audience and this in turn introduces a larger element of commercial pressure on the breakers to operate to improved standards—both owners and administrations will have to agree on the selection of breakers (and their proposed shipbreaking plans) in a very competitive international market. To this extent, the resolution of disputes relating to the operations of individual breakers may be settled more by simple commercial choice than recourse to international tribunals etc.

[27] To date, India has been successful in its subjective interpretation of the subject of asbestos—the new definition of hazardous materials '*contained in a ship*' (Regulation 20) may tighten the scope for such an interpretation.

6.3 Introduction of the Convention (and Amendments)

Requirements for the entry into force of the Convention are somewhat more complex than the usual IMO Convention. By Article 17, the Convention will enter into force 24 months after the date on which the following conditions have been met:

- 1.1 not less than 15 States have either signed it without reservation as to ratification, acceptance or approval, or have deposited the requisite instrument of ratification, acceptance, approval or accession in accordance with Article 16 (on signature, ratification, acceptance, approval and accession).
- 1.2 the combined merchant fleets of the States mentioned in paragraph 1.1 constitute not less than 40 per cent of the gross tonnage of the world's merchant shipping.

So far, these requirements are somewhat standard for IMO treaties, but in order to achieve the participation of the shipbreaking nations, requirement 1.3 also requires that:

- 1.3 the combined maximum annual ship recycling volume of the States mentioned in paragraph 1.1 during the preceding 10 years constitutes not less than 3 per cent of the gross tonnage of the combined merchant shipping of the same States.

Without the participation of at least some of the major shipbreaking states, not only would the Convention basically be ineffective, the whole *raison d'être* behind the Convention—namely a positive change to the current practices and procedures employed at the breakers—would be obviated.

These compound requirements, however, turn the criteria into something of a moveable feast since the numbers involved can change on an annual basis. Events such as the suspension of shipbreaking in Bangladesh during 2010–2011 could have an impact upon the criteria, but perhaps not too significantly, if taken over a period of 10 years.

Historically, the IMO has set a 5 year target from approval to final ratification for conventions but in the hope of an early introduction, this was reduced to just 3 years for the *HKC*.[28] Current assessment of dates for the Convention to enter into force can extend as far as 2020[29]; the earliest that many expect is 2015, which requires that the appropriate ratifications need to be completed in 2013. On 26 June 2013, Norway became the first (and to date—October 2013–the only) state to accede to the Convention. There are, however, mixed reports concerning the response of the ship breakers of the sub-continent to the new Convention with some suggestions of the breakers of India, Bangladesh and Pakistan joining forces in opposition to the Convention.

---

[28] Wallis (2009a), p. 1.

[29] The year 2020 was quoted as the expected year for the *HKC* to come into force in the EC's Impact report. European Commission (2012), p. 32.

6.4 Lacunae

Although the new Convention may appear to have a rather comprehensive scope, there are a number of apparent lacunae that have been identified. Whilst some of these lacunae may appear to be basic omissions or flaws, they do need to be appraised in the light of the basic practicalities and problems of disposing of surplus ships. The question of ships and recycling facilities not falling under the Convention (*i.e.* not belonging to Parties who are a signatory to the Convention) has been discussed above.

6.4.1 *Beaching*

In one of his many presentations on the Convention,[30] Mikelis, the former IMO facilitator for the Convention, considers (and dismisses) two of the more frequently cited shortcomings, namely absence of a ban on beaching and the need for pre-cleaning within the text. With regard to beaching, opponents of the Convention proffer that it is this very act of demolishing a ship at the shoreline that permits the current operations and procedures which result in the extensive release of hazardous materials and emissions into the surrounding environment. The frontpiece illustration of an aerial view of part of the Alang site shows dark trails of liquid wastes—presumably oils—running down the beach to the water's edge and creating a sheen over much of the sea in the image. At the diplomatic conference in Hong Kong where the Convention was adopted, of the 64 countries participating, only Ghana supported a ban on beaching[31] although opposition to beaching was held to be a major issue to other countries such as the Netherlands.[32]

However, an immediate and outright ban on beaching would remove some three quarters of the world's shipbreaking capacity, which is based upon the beaching method. It is also highly unlikely that those countries which follow such practices—India, Bangladesh and Pakistan—would ever vote in favour of such a restriction, since it would effectively outlaw their own operations. The acceptance of sufficient states which contribute shipbreaking capacity is the third of the three criteria necessary for the Convention to come into force, hence it would appear that ratification of the Convention by India and/or Bangladesh is a necessary precondition for the Convention to become effective; China is said to be already predisposed towards the Convention. Whilst this may demonstrate some necessity (albeit a most reluctant necessity) for beaching to continue, at least for some initial period, it does in consequence highlight the necessity for strict control of the handling of hazardous wastes and the promotion of high standards of environmental and health and

[30] Mikelis (2010b).

[31] Wallis (2009b), p. 2.

[32] SYBAss (2011).

safety protection on the beaches, a responsibility seemingly left to the operators of the shipbreaking sites. In this regard, Mikelis concludes that standards overall are expected to improve to meet the requirements of the Convention's guidelines, whilst countries practicing the beaching method will operate to the standards of the non-beaching shipbreaking states—although how ships broken up by hand on the beach will be dismantled with the same control as ships broken up alongside and with mechanical equipment is not spelled out. This is not exactly defined in the provisions of the Convention; the establishment and enforcement of operating standards by the shipbreaking states is, and will remain, a matter of national determination which, as has been demonstrated, may be quite flexible (if not erratic) to suit the needs of individual situations.

Notwithstanding the above, however, Regulation 15 of the Convention requires that Parties ensure that their ship recycling facilities are '*designed, constructed and operated in a safe and environmentally sound manner in accordance with the regulations of this Convention.*' This requirement for design and construction can hardly be applied to a beach/intertidal zone, the idea of 'constructed' focussing not only on the infrastructure for containment of wastes but also on management systems, including monitoring.[33] By remaining silent on the practice of beaching, the Convention, at least initially, is effectively helping to perpetuate the hazardous procedures of the sub-continent.

However, at a meeting of the Environment Committee of the European Parliament on 26 March 2013 (which also voted in favour of the introduction of a levy on all ships calling at EU ports to fund environmentally sound recycling), it was decided that the EU should go beyond the provisions of the *HKC* and unilaterally ban the beaching of ships for scrapping. Owners of EU ships that are sold and subsequently sent, within a 12 months period after the sale for scrapping on a beach or to a facility that is not on the EU list of approved breakers would be subject to penalties.[34]

6.4.2 *Pre-Cleaning of Hazardous Wastes*

Another major complaint of opponents of the current regime and the provisions of the Convention, at least in its initial format, relates to the fact that most ships arrive at the breakers containing a high tonnage of hazardous materials as cargo residues, operational wastes and materials incorporated into the actual structure and fittings of ships. Arguments usually require that such materials be removed from the ships by the owners who have benefitted from the operational life of the vessels over the years, rather than leaving such disposals to the shipbreakers (*i.e.* the site operators) who, in many cases, appear to have shown little regard for the consequences of effectively uncontrolled waste handling to date. Even today, when shipbreakers of

[33] Uytendaal (2012).

[34] Keating (2013).

the sub-continent may sample and identify substances considered potentially hazardous (especially asbestos), their removal is not necessarily undertaken in highly controlled conditions such as might obtain in Western states, but is undertaken in a manner that still allows the extensive production of airborne particles.

The effective removal of all hazardous materials can be a hugely expensive exercise, witnessed by the time and monies spent on the decontamination of the former British and New Zealand frigates *HMS Scylla* and *HMNZS Wellington* as dive sites in 2004 and 2005 respectively and the former US aircraft carrier *USS Oriskany* in 2006. The difficulty was also demonstrated by the numerous and often unsuccessful attempts to remove hazardous materials from the former French aircraft carrier *Le Clémenceau*, which consequently spent many months under tow seeking an accommodating haven. Such expenses are not willingly borne by the majority of ship owners, who expect a quick return for their surplus assets from the cash buyers. Those who are willing to internalise the cost—and thereby accept the principle of 'polluter pays'—usually do so in the acceptance of reduced scrap values paid by the breakers of some Chinese and Turkish yards, thereby demonstrating that financial returns and social and/or environmental responsibility need not be totally disparate. The final development and adoption of the Convention was nevertheless carried out during a period of distinct downturn for the shipping industry and measures that might add further to the costs of shipowners trying to dispose of uneconomic assets were not welcome.[35]

The real difficulty with regard to full pre-cleaning of a vessel prior to its final departure to the breakers is a practical matter, and lies in the form of the residual structure that a comprehensive removal of hazardous materials can present. The complete removal of insulation materials (especially in the engine and machinery spaces) and electrical cables, etc., can render a vessel not only unable to proceed under her own power, but also unseaworthy. Such a state would consequently make insurance for the voyage unobtainable—although a number of ships may proceed without insurance or even registration (which is a prerequisite for insurance)—but, more importantly, would impose a major cost of towing if indeed sufficient tugs of ocean-going capabilities were still available.[36] The consequences of this in turn would most likely result in a huge decline in the number of ships arriving at the breakers in China and the sub-continent, and could even result in the emergence of the industry in new locations.

The Convention does not restrict the locations at which pre-cleaning may take place, merely the locations at which demolition is undertaken. Regulation 8.2 states that ships destined to be recycled shall:

---

[35] Bradsher (2009).

[36] This is not to say that such voyages are not possible—witness the cases of the *Blue Lady*, *Le Clémenceau* and *Platinum II*, all huge ships that arrived at Alang under tow. These, however, are selected examples, the vast majority of ships for disposal still arriving at the breakers under their own power.

> conduct operations in the period prior to entering the Ship Recycling Facility in order to minimize the amount of cargo residues, remaining fuel oil and wastes remaining on board.

There is no specific reference to the further removal of other hazardous materials on board. Those carrying out shipbreaking activities are required to have the facilities and procedures to enable safe operations, and if the receiving breakers are deemed by the Competent Authorities not to be able to give such assurances then the cleaning may be carried out at other facilities which are adequately organised. By this process, the original NGOs demand that ships be pre-cleaned by their owners *prior* to despatch to the breakers appears to have transmuted to a situation whereby pre-cleaning is actually done *at* the breakers (or in proximity just prior to arrival, if the breaker does not have the appropriate facilities). In fact the pre-cleaning or stripping of the hazardous materials from a ship may be regarded as an initial, but integral, part of the shipbreaking process, and indeed some pre-cleaning, for example of the engine room spaces, may still be under way once actual demolition has begun.[37]

Appendix 5 to the Convention contains a Form of the Authorization of Ship Recycling Facilities. In the table of authorizations relating to the Safe and Environmentally Sound Management of Hazardous Materials, Yes/No columns cover the activities of removal, storage and processing[38] of hazardous materials. If a facility does not have authorization to *process* such materials, then the Ship Recycling Plan should contain details of where these operations are to be carried out. If the facility does not have authorization for the *removal* of hazardous materials, there is no reference as to where such activities are to take place. In the absence of such detail, it might be presumed that they can be carried out in any other appropriate location and arguments above suggest that removal is carried out at a location not too distant from the breaking site. However such locations may not necessarily be within the scope of the *HKC*, but should then remain within the remit of the *Basel Convention*.

Fundamentally, opponents of the Convention consider that the protection offered by *Basel* in the form of Prior Informed Consent, even considering the interstate acceptance of ship recycling plans, the right of port states to prevent exports and imports and the minimisation of transboundary movements of wastes etc., are all absent from the *HKC*. The subsequent handling of wastes beyond the shipbreaking sites is not considered by the *HKC*. The Convention is the first IMO convention to cover land-based activities, but its scope is restricted to the demolition sites only. The actual recycling operation for the scrap generated takes place at the various rolling mills etc. that are located elsewhere and excluded from the scope of the Convention. Since the Convention is a legal instrument and legal instruments are

[37] At this stage, the relevant asbestos-containing equipment in the engine room may be wrapped and identified as to its hazardous properties prior to removal, yet the engine room is probably fully open to the air.

[38] 'Processing' covers the incineration of hazardous materials, the reclamation of hazardous materials and the treatment of oily residues. Appendix 5, section 2.2 note 1.

based upon definitions, a more apposite title may perhaps be the Shipbreaking or Ship Demolition Convention rather than Ship Recycling Convention (which for the industry is obviously somewhat more anodyne).

6.4.3 Downstream Waste Handling

As mentioned above, the Convention contains no provisions for the handling and disposal of hazardous wastes emanating from ship demolition and this is considered a major shortcoming. This is especially the case since it has been this uncontrolled and un-enforced handling that has historically produced the legacy of pollution in the surrounding ecosystems and contributed significantly to many of the health problems suffered by both those directly involved in the breaking process, as well as others in the surrounding communities. Many photographs of the countryside around, for example, Alang show the accumulated wastes that have been disposed—*i.e.* dumped—by the breakers over time. The absence of control measures may leave that subject within the sphere of *Basel*, the observation of whose content and spirit to date has not been particularly well observed by some states, and once the transportation element of hazardous wastes has been removed by the *HKC*, subsequent waste handling is more likely to remain simply a measure of national rather than international responsibility.

6.4.4 Funding

Leading on from the question of upgrading facilities is the cost of undertaking such measures. It would appear from the Convention that the major impacts of change will fall upon the shipbreaking yard operators. Whilst the operating ship owners have the onus of compiling and maintaining IHMs, this practice had already begun with a number of owners before the adoption of the Convention with regard to new builds.[39] To some, the requirements of numerous inspections and certificates may appear little more than a paper chase, although the compilation of data for the IHM can represent a huge effort in the accumulation of information from a vast range of suppliers. It is at the actual beach sites, however, that the main improvements will have to be demonstrated; it is, after all, the poor legacy of bad practices and extensive pollution that was the initiator of the new Convention. The cost of imposing and maintaining upgraded and acceptable procedures will fall upon the

[39] Although the provision of IHMs applies only to commercial shipping under the *HKC*, a number of recent disposals of ex-Royal Navy ships were despatched to the breakers with full IHMs. A copy of the IHM for the submarine HMS *Conqueror* appears on the Ministry of Defence website. Ministry of Defence (2011a).

operator of those yards, which usually are operated as private family concerns rather than public companies.

The Convention, however, contains no provisions for the generation or supply of funds for the development of such sites. Funds are needed not only for the provision of suitable equipment, waste handling and storage facilities, training and medical facilities etc., but also for the cleaning of accumulated detritus and pollutions, although the soft nature of many beaches where demolition takes place may preclude the employment of large pieces of mechanical equipment. The provision of funding may also act as an incentive to ratification, whilst on the other hand, the cost of compliance may also promote a move away from what has been described as a cottage (albeit a very large cottage) industry in family hands. It has been suggested that an increase in costs of cash buyers—who albeit for a short time become ship owners—may actually promote a move towards further integration of the two parties.[40]

It was suggested by more than one interviewee that, should any funding to assist shipbreaking actually be forthcoming, then those funds were better directed towards the shipowners rather than the shipbreakers in order to encourage shipowners to despatch their obsolete vessels to the more suitable and compliant yards. The argument proffered against funding directly to breakers is that there are no guarantees that this would be used exclusively for operational improvements—improvements at the Turkish yards were met by funding emanating from the operators themselves.[41] This is actually in line with aspects of proposals examined on behalf of the European Commission for a funding scheme that would act as a catalyst and an interim scheme, pending the entry into force of the new Convention. It must be re-emphasised here that EU policy with regard to the development of shipbreaking on environmentally sound procedures and principles was not an operation that the EU was seeking to attract back from today's major centres and re-develop within the EU itself. The essential contribution that shipbreaking makes to those economies was acknowledged and stated clearly in the *Green Paper* of 2007[42] and reiterated in subsequent reports on the feasibility of creating and operating lists of ships for dismantling[43] and the development of a ship dismantling fund.[44] Instead, the EU is seeking to ensure that end-of-life ships that operate under a Member flag are demolished in an appropriate manner. Both of the issues (the creation of lists and a dismantling fund) are seen as measures that the EU might take to prepare for and encourage the early ratification of the *HKC*, but since they may also operate without the Convention, *i.e.* as measures aimed at enforcing sustainable shipbreaking in their own right, they are discussed more fully in the following chapter, together with the EU ad-hoc Regulation.

---

[40] Drury (2009).

[41] Comments made by interviewees in June 2010.

[42] *Green paper on better ship dismantling*. COM(2007) 269 final. Brussels, 22.5.2007.

[43] Mudgal et al. (2010).

[44] Milieu Ltd. & COWI (2009).

In November 2011, an MoU was signed between Bangladesh and the Norwegian Agency for Development Co-operation (NORAD), the latter providing a US$5–6 million fund for the provision of safe and environmentally sound ship recycling facilities in Bangladesh, NORAD requesting that the IMO manage the project.[45] However, this practical initiative was criticised by the NGO Shipbreaking Platform as a non-transparent, society-inclusive move that may '*simply institutionalise beaching and the importation of toxic ships.*'[46]

6.4.5 *Scope of Vessels*

Whilst the *HKC* applies to ships entitled to fly the flag of a Party or operating under its authority, it does not apply to:

- Ships of less than 500GT
- Ships operating throughout their life only in waters subject to the sovereignty of the state whose flag the ship is entitled to fly
- Warships, naval auxiliaries, or other ships owned or operated by a Party and used, for the time being, only on government non-commercial service.[47]

Excluding those vessels to whom the Convention does not apply, the size of the world's merchant fleet has been estimated at between 42,000 and 45,000.[48] Ships below 500GT will include many small vessels such as fishing boats and other non-ocean going ships; although their percentage of overall GT may only amount to some 1 %, collectively they account for some 36 % in numbers of vessels that are scrapped each year.[49] The French monitoring organisation Robin des Bois reported in 2008 on the large number of fishing boats that were being scrapped as their owners opted for the compensation that was being offered by France and by the EU to reduce the size of the fleet in line with European fishing policy, the boats being rapidly demolished by mechanical excavators, with little or no regard paid to the separation of hazardous materials or the protection the surrounding environment.[50] Since the smaller vessels may incorporate a high percentage of composite material rather than recyclable materials, their disposal may actually represent a financial

[45] Challenges envisaged by NORAD include the need to establish a real commitment in Bangladesh from both the Government and the yard owners themselves, together with the need to change existing responsibilities and policies for an industry in which many ministries are involved. Oftedal (2012).

[46] NGO Shipbreaking Platform (2011).

[47] Article 3 Application.

[48] Chang et al. (2010).

[49] COWI (2011), p. 6.

[50] Robin des Bois (2008).

liability to their owners[51]; nevertheless, although each unit may be relatively small, the combined weight of hazardous materials that they account for may be high.

The exclusion of all these small vessels (including small vessels confined to coastal waters) may be due more to a question of logistics, not so much of the vessels themselves, but of the vast number of small yards scattered around many countries of the world, these yards together forming a huge burden of the compilation and verification of IHMs, certifications and inspections that would probably prove extremely difficult to manage for most states and for the ship owners. The cost of moving small vessels any great distance for demolition is also likely to be excessive in relation to the returns that may be made on the resultant scrap and attempts to enforce such requirements might simply lead to a high level of abandonment. Nevertheless, the impact of these small vessels taken together may be quite significant although the problems to which they might give rise are understandable (at least in the initial stages of the Convention). The question remains, however, that, if the 500GT limit is amended, exactly where the line should be drawn, since even sailing dinghies fall within the Convention's definition of 'ship.'

The exclusion of naval ships and government vessels on non-commercial service is, however, a quite distinct and different matter, exempting governments, as it does, from the obligations placed upon commercial owners. Although such vessels (which individually may be very large) fall outside the scope of the Convention, Article 3 also requires that:

> ..each Party shall ensure, by the adoption of appropriate measures not impairing operations or operational capabilities of such ships owned or operated by it, that such ships act in a manner consistent with this Convention, so far as is practicable.

The new international standard ISO30000 on ship recycling facilities[52] merely refers to 'ships' and does not distinguish between naval and merchant vessels. Organisations such as INTERTANKO and the ICS support the inclusion of warships and other government owned vessels, the latter considering such a move as '*a commendable undertaking displaying goodwill on the part of the Community*,' but consider that an extension to include vessels of less than 500GT as presenting complexities that might hinder the speedy entry into force of the Convention.[53] Although the number demolished each year is relatively small, warships—even with their relatively high proportion of more valuable metals—can represent a challenge for even the most conscientious of breakers that goes beyond that associated with merchant ships. These ships, with their extensive compartmentalisation plus wide array of inbuilt equipment, can present a plethora of hazardous materials in both the insulating and electrical gear. In addition, the relatively long life of many naval vessels in comparison to the mercantile fleet can result in a much higher presence of hazardous materials within their structures. The results of the

[51] COWI (2011), p. 7.

[52] International Organisation for Standardisation (2009).

[53] INTERTANKO and ICS responses to the EC public consultation on the 2007 Green Paper. European Commission (2009).

recent UK Government's public consultation on ways to render the growing fleet of decommissioned Royal Navy nuclear submarines[54] free of all nuclear and contaminated materials, prior to the demolition of the remaining sections of the hull at commercial UK breakers, will still leave the final disposal of the remnants of these vessels outside the provisions of the Convention. The Swedish MEP Carl Schlyter, who has been steering the new Ship Recycling Regulation proposal—see Chap. 7—through Brussels also called for the exemption of naval vessels to be lifted.

It was the reports in the *Baltimore Sun* of the dreadful conditions under which warships were being demolished in the shipbreaking yards of America and India that began the investigations and subsequent concern over conditions in the industry in 1997. It was also somewhat ironic that when in September 2008 the former Royal Navy landing ship *HMS Intrepid* was despatched to Liverpool for scrapping, the workers at the Canada Dry Dock took industrial action to demonstrate against the perceived level of hazardous materials aboard and the lack of consultation over Health and Safety protection.[55]

6.4.6 The Impact of the New Administrative Requirements

The impact that the prescribed process of inspections, approvals and documentation will have upon current practices is, as yet, unclear. Final (operating) owners are unlikely to embark upon such a regime if ships can be sold simply to cash buyers, whilst the change of flag that usually accompanies such sales would merely break the link of documentation between the (original) flag state and the shipbreaking state. Cash buyers themselves might be loath to initiate such a process of documentation, which could add significantly to the disposal time and to the cost of finance,[56] yet even if such a process were to be undertaken, it is questionable whether a number of flag states actually have the resources (or even the will) of a maritime authority able to provide such services, many states having relinquished the right to operate their register to third parties simply as a means of raising revenue.

The outcome of this administrative regime is yet to be determined. On the one hand, there may be complete compliance with the requirements of the Convention—both in the writing and in the spirit; on the other hand, the weight of the disciplines imposed may simply result in non-compliance, non-ratification and/or the emergence of new centres of shipbreaking in areas not currently associated with the industry.

[54] Ministry of Defence (2011b).

[55] Unite the Union (2008).

[56] Cash buyers speaking at the 2012 Ship Recycling Conference were at pains to dispel the idea that there was a high profit margin associated with their disposal activities.

In 2011, the European Commission commissioned from COWI a guideline report on the recovery of obsolete vessels.[57] This report outlines a three component strategy for the recycling of such ships, the first of which consists of a policy level—establishing a policy that underpins a commitment to environmentally sound recycling, and a practical level—establishing appropriate procedures to actually identify all government owned ships. This is followed by a series of measures deemed necessary to prepare a ship for disposal, including an IHM and ensuring that a ship is seaworthy for its final voyage or tow, and finally a list of requirements to ensure that the recycling facility meets all required standards. Special consideration may be required for vessels which do not have a steel hull, for example mine countermeasures ships, which may have a hull of glass reinforced plastic and hence of very little residual scrap value.[58]

Now that the final guidelines of the Convention have been agreed, it is more likely that the rate of signing and ratifying of the Convention will increase, but there remains the question of national interpretation of these guidelines in the face of national interests, as the sometimes conflicting judgements from the apex courts of the shipbreaking states in the sub-continent have demonstrated. A tightening of standards and legislative enforcement has, in the past, merely served the interests of competitive states, witness the general switch of end-of-life tankers from the Indian to the Bangladeshi yards once India began to enforce its gas-free policy. There remains the possibility that a general move by all the established states towards the standards envisaged by the Convention could simply result in the development of a second level market amongst states not yet fully identified with shipbreaking activities.

6.5 Equivalence with Basel

At the 9th Conference of the Parties to the Basel Convention held in Bali in 2008,[59] Decision IX/30 called upon its Open-ended Working Group (OEWG) to carry out a preliminary assessment as to the equivalence of control and enforcement levels that existed between the *Basel* and *Hong Kong Conventions*. The results were presented to the following session of COP10 in 2011 in the Annex to a Note by the Secretariat[60]; some of the main issues are addressed here, including some which have already been referred to in the above section.

In terms of coverage, *Basel* regards all ships for disposal as hazardous waste, without distinction as to size, operation or ownership, whilst the *HKC* is specific as

[57] COWI (2011).

[58] COWI (2011), p. 92.

[59] Basel Convention, Ninth Conference of the Parties (COP9) of the Basel Convention, 23–27 June 2008, Bali, Indonesia.

[60] Basel Convention Secretariat (2011).

to the exceptions to scope—government-owned vessels etc. The scope for materials coverage, whilst overlapping, also shows distinctions between the two conventions, *Basel* listing specific materials and wastes, the *HKC* being applicable to any material that might be deemed hazardous to both human and environmental health. On the other hand, the *HKC* does not cover certain materials detailed within the *Basel technical guidelines on ship recycling*.

Basel requires that the generation and export of hazardous materials be kept to a minimum, with wastes treated in the producing state (the proximity principle), or if exported, to be treated only at facilities where sound environmental management is practiced. Control under the *HKC* has a somewhat similar requirement for minimisation beginning with the design, building, maintenance and operation of a ship, together with controls for the standards of environmentally sound management to be observed within the shipbreaking facilities. Where the two diverge is in the post-dismantling handling of the materials, the *HKC's* provisions ending at the recycling facility and excluding any handling and disposal of materials thereafter. Whilst the *HKC* may therefore cover the life cycle of a ship, the same cannot be said for the processes required in the dismantling state. *HKC* also splits the authorization and certification processes, giving responsibility for the surveying and certification of a ship up to the point of it being ready for recycling to the flag state, and responsibility for recycling facilities to their appropriate recycling states; the provisions under *Basel* are not as specific. *Basel*, on the other hand, is somewhat more emphatic on the question of punishing conduct that is in contravention of the Convention and declaring such trading as illegal activity (although no specific penalties are defined).[61]

In 2011, however, it was announced by the EU that their examinations of the equivalence issue had proved positive and that it considered that the *HKC* did indeed prove to have equivalence in as much as it applied to end-of-life ships. Although the decision on equivalence by the Parties to Basel at COP10 was by no means unanimous, they encouraged an early entry into force of the *HKC*, whilst acknowledging that the Basel Convention should continue to assist countries to apply *Basel* as it related to ships.[62] This effectively drew a line under any official resistance to the *HKC* and acted as a spur to its ratification, the EC, as one might have come to expect, moving ahead with some zeal, not only to add its weight behind the IMO Convention, but looking to go beyond its existing provisions.

---

[61] *Basel Convention*, Article 9.

[62] Decision BC-10/17, Parties to the Basel Convention, COP10, October 2011. A number of states at COP10 had been actively involved in the development of the *Hong Kong Convention* and hence supported the idea of its equivalence in control and enforcement. Those who had not been so involved did not support the degree of equivalence; MEPC 63 of the IMO therefore requested that the maritime administrations of Member States should brief their counterparts in the environmental ministries as to the extent to which the new mandatory requirements were practicable and which would generate a level of control and enforcement equivalent to Basel. International Maritime Organisation (2012).

Details of the resultant new ad-hoc Regulation developed by the EC to spur on ratification of the *HKC*, not only within its own Member States but indirectly amongst the IMO membership generally, is covered in the following chapter. The acceptance of the *HKC* by at least some of the Basel Parties has perhaps placed the Basel Action Network in somewhat of a difficult position, now standing as it does between the two Conventions.

6.6 Interim and Preparatory Measures

Until the *HKC* actually enters into force, a voluntary implementation of the relevant technical requirements of the Convention at the earliest opportunity is urged by its promoters. At the Hong Kong diplomatic Conference 2009, states including the recycling nations were also invited by Resolution 5 to adopt such measures on a voluntary basis. Amongst those technical requirements, Mikelis (IMO) identified a number that might be considered as suitable interim measures, including:

- The provision of an Inventory of Hazardous Materials for ships going to recycling
- The preparation of a Ship Recycling Plan for ships going for recycling
- The obligations of both shipowners and ship recycling facilities with regard to safe-for-entry and safe-for-hot work provisions
- Compliance of the Convention's safety, health and environmental standards by the recycling facilities.

India and Turkey already have requirements for the preparation of a Ship Recycling Plan and an agreement on this and other measures between the major shipbreaking associations; the arrival of ships at the facilities in a clean and gas-free condition and other measures could effectively bring them into force.

The early adoption of technical requirements has also been undertaken by a number of shipowners' associations, including the ICS, BIMCO, INTERTANKO and INTERCARGO. Published by the Maritime International Secretariat, the collective guidelines *Selling Ships for Recycling*[63] begin by urging both shipowners and ship recyclers to recognise their responsibilities to environmental protection and to the health and safety of workers in the yards and begin implementing the requirements of the Convention during the transitional period in order to be ready for compliance once the Convention enters into force. It adds that, during the interim period, standards of quality in the various yards may be expected to vary, but its provisions will increasingly be required by flag states and an adherence to the interim measures should be regarded as a sign of good faith.

[63] Maritime International Secretariat (2009).

As a practical indicator of the improving conditions at some yards, it was further suggested that a socially responsible shipowner such as Maersk, who has an established practice of sending its ships for demolition under controlled conditions at specific Chinese yards, should send some of its ships to such yards in the sub-continent as could demonstrate positive improvements in their procedures, this to act as an incentive to other shipowners. Any payments received for the ships beyond the level that might be obtained from Chinese yards might be put to a fund for promoting the provision of an improved infrastructure for the management of hazardous wastes.[64] The Maersk response was that clear evidence of improvement, and over time, would first be required.

Another improvement that would add to the general enhancement relates to the role of the cash buyers. It was suggested at an EMSA workshop on ship recycling held in Lisbon in 2011[65] that this small but increasingly important group needed to adopt and pay heed to the requirements of the Convention with regard to documentation etc., rather than merely focus upon maximising financial returns.

At the end of 2010, the cash buyer Global Marketing System (GMS) announced, in co-operation with the classification society Germanischer Lloyd (GL), that it would apply the principles of the *HKC* to green ship recycling, whilst proving that both GMS and the recycling sites selected comply with the requirements of the Convention, this service being for ship owners who, in advance of the Convention coming into force, look to have their vessels '*responsibly recycled*.'[66] GL describes itself as the only classification society which continuously builds up an international network of GL-approved 'hazmat experts' (a somewhat self-sustaining claim) in order to ensure compliance with the requirements of both the IMO and ISO 30005 when compiling IHMs. In fact the preparation of IHMs and allied activities is becoming a popular new service on offer by classification and related organisations in the face of the oncoming Convention.[67]

Shortly after the adoption of the Convention, in February 2010 an agreement was made between the government of Japan and the Gujarat Maritime Board that would help bring shipbreaking at Alang up towards the required international standards; the resultant Memorandum of Agreement centred upon both financial assistance and the transfer of technology to India.[68]

---

[64] McMahon (2011), p. 3.

[65] European Maritime Safety Agency (2011).

[66] The Maritime Executive (2010).

[67] Assistance with the compilation of IHMs is also being offered by the classification societies DNV, RINA, ABS, Lloyds Register and Class NK, Japan. Here the offer is usually of assistance in the drafting, rather than final production of the IHM, as well as auditing and certification.

[68] Diligent Media Corporation (2010).

References

Basel Convention Secretariat (2011) Environmentally Sound Recycling of Ships. Note by the Secretariat UNEP/CHW.10/18. Conference of the Parties to the Basel Convention, tenth meeting, Cartagena, Columbia 17–21 October 2011

Bradsher K (2009) Agreement on ship recycling wins wide support. BAN Toxic Trade News. 14.5.2009 www.ban.org/BANNEWS/2009/090514/agreement_wins_wide_support.html accessed: 3 January 2010

Chang YC, Wang N, Durak OS (2010) Ship recycling and marine pollution. MPB 60:1390

COWI (2011) Recovery of obsolete vessels not used in the fishing industry. Report P-74494-B1-Final. For the European Commission, DG Environment. December 2011 Kongens Lyngby, Denmark.

Diligent Media Corporation (2010) Japan's expertise to refurnish Alang. www.findarticles.com/p/news-articles/dna-daily-news-analysis-mumbai/mi_811/. Accessed 13 Apr 2010

Drury S (2009) United Kingdom: how does the future IMO Ship Recycling Convention address the issue of the cash buyer? Tradewinds Ship Recycling Forum, 19 January 2009, London

European Commission (2009) An EU strategy for better ship dismantling. Communication {SEC (2008) 2846} COM (747) 2008

European Council (2012) Commission staff working document. Impact assessment accompanying the document Proposal for a Regulation of the European Parliament and of the Council on ship recycling. SWD(2012) 47 Final. 23 February 2012, Brussels

European Maritime Safety Agency (2011) European Maritime Safety Agency Ship Recycling Workshop, 27–28 June 2011, Lisbon

Galley M (2013) Hazardous materials in shipbreaking – where do the liabilities lie? PhD research thesis, Southampton Solent University, Southampton

Harrison J (2009) Current legal developments. International Maritime Organisation. Int J Marine Coastal Law (2009) 24:727–736

International Maritime Organisation (2012) Report of the Marine Environment Protection Committee on its Sixty-third session. IMO MEPC 63, 23 14 March 2012

International Organisation for Standardisation (2009) ISO30000:2009 Ships and marine technology – ship recycling management systems – specifications for safe and environmentally sound ship recycling facilities. ISO, Geneva

Keating D (2013) MEPs Call for ship recycling fund. European Voice. www.Europeanvoice.com/meps-call-for-ship-recycling-fund26.3.2013. Accessed 27 Mar 2013

Maritime International Secretariat (2009) Selling ships for recycling. Guidelines on transitional measures for shipowners in preparation for the entry into force of the IMO Hong Kong International Convention for the Safe and Environmentally Sound Recycling of Ships

Martinsen K (2009) Environmentally sound ship recycling. DNV Bulk Carrier Update – No.1. Supplement Fairplay, 30 April 2009, London

McMahon L (2011) Maersk shuns call for ship recycling in South Asia. Lloyd's List, 31 October 2011, London

Mikelis N (2010a) Introduction to the Hong Kong Convention and its requirements. In: Ship Recycling Technology & Knowledge Transfer Workshop, 14–16 July 2010, Izmir

Mikelis N (2010b) The Hong Kong International Convention for the Safe and Environmentally Sound Recycling of Ships. Multi-year Expert Meeting on Transport and Trade Facilitation, 9 December 2010, Geneva

Mikelis N (2011) The Hong Kong Convention on Ship Recycling. In: European Maritime Safety Agency Ship Recycling workshop, Lisbon, 27–28 June 2011

Milieu Ltd. & COWI (2009) Study in relationship to options for new initiatives regarding dismantling of ships. Executive summary. Brussels, August 2009

Ministry of Defence (2011a) Green passport HMS Conqueror. www.MOD.org/NR/rdonlyres/OBD9AB5C-8471-6EB6FAD544C/6EB6FAD544FC/0/20111018.ConquerorGreenPassport.pdf

Ministry of Defence (2011b) Submarine dismantling project (SDP) Consultation document. MOD, 28 October 2011
Mudgal S, Benito P, Kong MA, Dias D, Carreno AM (2010) The feasibility of a list of 'green and safe' ship dismantling facilities and of a list of ships likely to go for dismantling. Bio Intelligence Service Report for the European Commission (DG ENV) Final, 4 January 2010
NGO Shipbreaking Platform (2011) Environmental NGO calls on Norway to reform its shipbreaking aid packages for Bangladesh. NGO Shipbreaking Platform, 13 May 2011. http://recyclingships.blogspot.com/2011/05/environmental-ngo-calls-on-norway-to.html. Accessed 7 Feb 2012
Oftedal S (2012) Development of ship recycling facilities in Bangladesh – NORAD Project update. In: 7th annual ship recycling conference, 29–30 June 2012, London
Robin des Bois (2008) An Alang scheme. www.robindesbois.org/english/sea/alang_scheme08.html. Accessed 9 Dec 2011
Schneider S (2011) Compiling an inventory of hazardous substances. In: 6th annual ship recycling conference, 14–15 June 2011, London
SYBAss (2011) Clarification for SYBAss members regarding recycling convention. Final version, Superyacht Builders Association, 8 April 2011
Tanker Operator (2009) GL holds recycling seminar. Tanker Operator, 8 May 2009, London
The Maritime Executive (2010) GMS implements Hong Kong Convention. www.maritime-executive.com/pressrelease/germanischer-lloyd-gms-hong-kong-convention. Accessed 1 Feb 2012
Townsend R (2012) Successful initiatives for implementing the standards of the Hong Kong Convention so far. In: 7th annual ship recycling conference, 19–20 June 2012, London
Unite the Union (2008) Workers fear for safety over scrapping of HMS Intrepid in Liverpool. www.unitetheunion.org/news_events/archived_news_releases/2008archived. Accessed 29 Jan 2012
Uytendaal A (2012) How can a beach come up to ISRA standards? In: 7th annual ship recycling conference, 19–20 June 2012, London
Wallis K (2009a) Green light for IMO recycling rules. Lloyd's List, 18 May 2009, London
Wallis K (2009b) IMO rejects campaigners' call for ban on beaching. Lloyd's List, 4 May 2009, London

Chapter 7
Other Initiatives

While the *HKC* proceeds through to eventually coming into force, other issues are being, or have been, independently pursued to encourage ratification of the Convention by various states within the IMO. New initiatives, some of which are examined below, may have a role in promoting the coming into force of the Convention or have a role to play in their own right, either alongside the Convention, or in place of the Convention, should the treaty not actually come into force.

7.1 The EU Response

7.1.1 Initial Reactions

The European Union appears ever anxious to promote such provisions as will facilitate sound environmental management and the handling and elimination of hazardous waste—witness the early incorporation of the *Basel* principles, including the *Basel Ban*, into its *WSR*, the acceleration of the phase-out of single-hulled tankers and the incorporation of the *IMO Convention on Anti-fouling Systems on Ships* into EU legislation before that Convention entered into force.[1] Its reputation for urging developments in the amelioration of the adverse effects of the shipbreaking industry has sometimes been to the discomfort of the IMO, which has to proceed more slowly through a process of consensus within its wider range of members, the EC calling for '*concrete regulatory action at the EU level which moves beyond the weak remedies of the IMO*.'[2]

On this basis, the Commission considered an early adoption of the *HKC* into European legislation before the Convention was formally ratified by Member

[1] *Regulation (EC) 782/2003 of the European Parliament and of the Council of 14 April 2003 on the prohibition of organotin compounds on ships.*

[2] *European Parliament resolution on the EU strategy for better ship dismantling, 26 March 2009.*

M. Galley, *Shipbreaking: Hazards and Liabilities*, DOI 10.1007/978-3-319-04699-0_7,

States, but initial lack of support meant that this idea had to be abandoned at the time. Nevertheless, the Commission asked that Members give a high priority to early ratification, which it refers to as '*a major achievement for the international community*.'[3] Interim measures and measures to promote the ratification of the Convention are examined below. Measures which were progressively considered by the Commission led eventually to the formulation of an ad-hoc Regulation—subsequently entitled the *Ship Recycling Regulation*. This is intended not only to promote the Convention, but includes measures that actually go beyond the *HKC* (even if only on an interim basis) that could stand in their own right as incentives for Members to dispose of ships in an environmentally sound manner. On this basis, actions by the EU have been included in this chapter with other independent initiatives as provisions that may be considered as running parallel to, or even in the absence of, the *HKC*.

Given the difficulties experienced by many campaigning NGOs and various national courts promoting the principles of *Basel* to the disposal of end-of-life ships, the absence of any specific measure relating to ships (within the Convention) has proved to be such that shipowners have found it quite a simple matter to bypass *Basel* and even the European Commission itself announced in June 2011 that it was unable to enforce the provisions or spirit of *Basel*. This admission was re-iterated in the report from the EMSA workshop held in Lisbon later that year, which highlighted the lack of adequate recycling capacity within OECD countries, together with the difficulty of identifying the state of export and the point at which the decision to discard is made. It was further offered that the export ban enshrined in the *WSR* may even be blocking any good initiatives that shipowners may devise to scrap their ships at sites outside the OECD which offer sound environmental practices.[4]

Options suggested by the IMO included either the enactment by the EU of measures based upon the standards and exempting end-of-life ships from the *WSR*, or voluntary measures which would allow compliant ships to be exempt from the *WSR*—in both cases, the role of the *WSR* was presented as an obvious hindrance. The response suggested by the EC was to adopt the Convention into EU law and allow ships covered by the Convention to be dismantled in facilities outside the OECD as long as those facilities were third party certified as being compliant with the *HKC* standards,[5] which was ultimately to be the path to be followed. A number of alternative scenarios were presented at the workshop for comment; the worst case scenario proposed was considered to be no changes at EU level and the Convention not entering into force.

An alternative and opposing viewpoint offered was that far from taking end-of-life ships out of the provisions of the *WSR*, clearer guidelines on exactly when a ship becomes waste might strengthen the Regulation; to do otherwise might simply offer

[3] European Commission (2010), p. 2.

[4] European Maritime Safety Agency (2011).

[5] European Maritime Safety Agency (2011).

further loopholes to shipowners. Further, to strengthen the *WSR* would require greater emphasis on pre-cleaning, ultimately involving partial dismantling—hence bringing the focus back to the practical problems associated with that activity prior to a ship departing to the breakers. The conclusion therefore was that a rational approach would be to concentrate efforts on promoting improvements in waste handling at the shipbreaking sites.[6] The corollary of this is the lack of incentives to expand recycling operations in the OECD countries themselves; this was considered to be due to the high level of competition for recycled materials in the sub-continent and the low wages/high prices for end-of-life ships that obtain there.

At the opening of an NGO Shipbreaking Platform exhibition of photographs of current shipbreaking operations in 2011, Potočnik, the European Commissioner for Environment, acknowledged that current legislation was intended to control the movement of waste, but its application to ships was difficult as they represented '*a kind of moving target*,'[7] prompting the need to:

> . . .think globally to solve the problem [and] think long term. . .to push out the sub-standard operators, and provide the incentives for the good operators.[8]

This, Potočnik added, was the objective of the *HKC*, and he laid the responsibility on shipowners to limit the use of hazardous materials in their vessels at all stages of their life and advise shipbreaking workers of the presence of hazardous material on board *and protect them from it* (although exactly how this protection would be effected was not spelled out).

The sentiments expressed above reflected the EU's recognition of the importance of shipbreaking, both to the economies of the breaking nations and to the reuse of resources, although the current performance of shipbreaking in South Asia was considered '*environmentally destructive and degrading to humankind*.'[9] These sentiments were also contained in two reports commissioned by the European Commission; in 2009, the EC (DG Environment) commissioned a study into the feasibility of a ship dismantling fund,[10] and the following year a report on compiling a list of 'green and safe' ship dismantling facilities and of ships likely to go for dismantling.[11]

[6] European Maritime Safety Agency (2011). It was generally agreed at the EMSA workshop that Turkey already had a regulatory regime similar to that of the EU and that China, whilst still having some ground to make up in terms of waste handling, was making good progress and was worthy of receiving ships from the EU states.

[7] Potočnik (2011).

[8] Potočnik (2011).

[9] European Parliament (2009) Preamble.

[10] Milieu Ltd. & COWI (2009).

[11] Mugdal et al. (2010).

7.1.2 European Funding Study 2009

In February 2005 Greenpeace published a research report *The Ship Recycling Fund – Financing environmentally sound scrapping and recycling of sea-going ships*, which had been compiled by the Dutch company Ecorys. The report considered a strategy for establishing a fund, operated under the auspices of some UN organization that would collect fees and disburse them to yards, which it would certificate and control, and which were able to undertake environmentally sound shipbreaking. The fund would also be used to finance research into improved ways of scrapping. The report considered the collection of fees at vessel registration, but concluded that this would put the onus solely on new builds rather than across the whole fleet. As a preferred option, it proposed the application of fees, to be collected by flag states, throughout the lifetime of a ship. This report was issued over 4 years prior to the adoption of the *HKC*.

Commissioned by the EC and issued in August 2009, the study[12] undertaken by the Belgian consultancies Milieu Ltd. and COWI considered the potential of an EU ship dismantling fund to assist the coming into force of the Convention and whether such a fund might in fact be operable on a global basis. The possibility of such a fund was outlined in the *Green Paper* of 2007. Following a public stakeholder consultation and a subsequent stakeholder workshop at the DG Environment in 2009, three options for a ship recycling fund were considered, recognising that the few sites that today offer shipbreaking in a manner that is deemed (relatively) safe and environmentally sound do so at prices per net ton to shipowners that are lower than those on offer from other yards.[13] Recognising also that the new Convention appears to have deficiencies in terms of enforcement, the study examined measures that might enhance the drive for early entry into force, reiterating the proposals expressed in the integrated maritime policy for the European Union, issued in 2007.[14]

Focussing on ships with strong ownership or flag links to the EU, the objective was to provide a fund that would give incentives to shipowners to scrap their end-of-life ships at yards displaying acceptable practices and bridge the gap that exists between prices offered there and by more conventional yards. A number of criteria were developed; in order to evoke the 'polluter pays' principle; a system that financed the fund via the shipping industry (or the consumers of shipping services) was considered more appropriate than financing by taxpayers. The fund must also avoid distortion of competition by subsidising cost-inefficient EU shipbreaking operations whilst also avoiding a weakening of the EU shipping industry. Recognising also the backlog of surplus ships resulting from adverse trading

[12] Milieu Ltd. & COWI (2009).

[13] The report estimated the extra costs arising from a more environmentally sound operation was between the wide range of US$25 and US$150 per LDT, with US$100 being the central figure. Milieu Ltd. & COWI (2009), p. 21.

[14] European Commission (2007).

conditions and the planned phase-out of single-hulled tankers that will require disposal in the interim, the fund would need to be accumulated quickly, which in turn suggests a start-up phase from government funding (in opposition to the previous criterion). The study also recognised that, given the global nature of shipping and the ease with which transfers between flags are made, a re-definition of EU responsibility might be based on the EU's share of world shipping overall; on this basis, it might be appropriate to consider the number of ship visits to ports worldwide and to the EU, and although the fund was initially considered as EU specific, it might later be expanded to a more global level.[15]

In terms of funding, it was determined that a revolving fund, which received income from levies or taxes, appeared preferable to an endowment or sinking fund, both of which required a large initial investment. The financing mechanism could arise from either an up-front charge at construction of a vessel; a recurrent tax on shipping; or a charge on shipping calling at EU ports. An up-front charge is deemed to be relatively simple to administer, but would apply only to new ships—thereby avoiding charges on existing ships and the 'polluter pays' principle, and may also lead to distortions between EU and non-EU ships, in turn promoting tax avoidance and flag changing. The fund would also take some time to accumulate. An annual charge would offer both a wider tax base and a faster growing fund, but could also lead to the same disadvantages as a charge on new builds. A charge on all ships calling at EU ports would provide a quick growing fund that could be administered as other port charges. Belgium's response to the public consultation on the Green Paper with regard to the funding sources was that it considered a charge on new builds, which are probably the more environmentally sound, would disregard all older, existing vessels, and that a charge on port visits may prove detrimental to the competitiveness of European ports. This left the annual tax on shipping as the most favourable of the options.

Whatever the method of accumulation, funds could be delivered either as compensation for EU ships that use approved recycling facilities; subsidizing environmentally sound facilities—either EU or non-EU—that comply with EU standards; or the use of a voucher system from certified recycling facilities, the vouchers being produced by the shipowner when requesting reimbursement. Given both the speed with which a fund should accumulate and the need to base it on the 'polluter pays' principle, the conclusion of the Milieu/COWI analysis was that should the EU determine that a ship dismantling fund was a necessity as an incentive for the Convention entering into force as early as possible, the appropriate format, as an interim measure, would be for an EU revolving fund, based on recurrent charges on ships calling at EU ports and disbursed on production by the shipowner of evidence of scrapping at an approved facility (the voucher scheme).

It was the conclusion of the report that, without such a fund, there would be little incentive in Europe to use approved and environmentally sound breakers, with the fund making good the shortfall in the lower prices to owners that such facilities

[15] Milieu Ltd. & COWI (2009), p. 4.

would be able to offer with their higher costs. The *HKC* makes no reference to the funding of improvements to current shipbreaking practices.

Throughout the report, neither is there any reference to the concept of pre-cleaning as an obligation of shipowners. By concluding that the option of directing owners towards yards is deemed to be safer and more environmentally sound, the implication is that the onus for improvements in shipbreaking practices lies with the breakers themselves, to be funded from a probable or possible increase in business directed their way.

On 26 March 2013, Members of the Environment Committee of the European Parliament voted to create a Europe-wide ship recycling fund, designed to '*finance environmentally sound ship recycling and internalise the costs of proper hazardous waste management*,' the fund to be raised from a fee of €0.03 per GT[16] to be levied on all ships calling at EU ports. The fund would be available for use by both EU and non-EU ships, with owners having the choice between an annual recycling levy (paid directly to the fund) or a fee per port call basis, paid to the relevant port authorities. By imposing a levy on actual port calls, the aim was to make it impossible for shipowners to evade the charges by reflagging to non-EU registers.[17] Reactions to the proposal vote were mixed, with the NGO Shipbuilding Platform in strong support and the port and shipowner representative groups registering their opposition to the move. At the same time, the Parliament's Environment Committee also voted to remove end-of-life ships from the *WSR*, a move that the NGO Shipbreaking Platform characterised as violating the EU's and Member States' obligations under *Basel* and the *Basel Ban*.[18]

A further report examining funding options was commissioned by the NGO Shipbreaking Platform from the Profundo economic research organization in Amsterdam and issued in January 2013. This report is discussed in Sect. 7.1.4 below.

7.1.3 European Listings Study 2010

In the 2010 report, part A was devoted to an examination of the possibility of developing criteria that would facilitate the compilation of a list of ships under EU flag or ownership that were *likely* to go for dismantling (emphasis added), such a provision being aimed at supporting the EU strategy on ship dismantling and increasing '*the effectiveness and enforcement of the WSR and the future HKC*.'[19] The difficulty of determining at what age a ship might normally be sent for demolition was acknowledged, and this in turn led to the development of a number

[16] Messenger (2013a).

[17] Messenger (2013b).

[18] Messenger (2013b).

[19] Mugdal et al. (2010), p. 5.

of criteria that together might help the various national authorities decide whether any vessel was a likely candidate for inclusion. These criteria were defined as:

- Legal obligations—*e.g.* the requirement to scrap single-hulled tankers.
- Age and type of ship—the older a ship, the more likely that earning abilities would decline and running costs increase.
- Flag—to indicate the effectiveness of the country's Port State Control and safety measures.
- Number and reasons for deficiencies and detainments—indicating maintenance level.
- Economic factors—a very important driver in the scrapping decision.
- Classification society—as a measure of effectiveness of its inspection performance.

Three scenarios were developed: Model A, based upon unfavourable economic conditions; Model B, for ships falling within the average dismantling age, and found to have been detained for certain deficiencies etc.; Model C, basically for ships flying the flags of specific targeted states. It is not intended to examine here in detail these various alternatives, which were set against estimated dismantling volumes through to the year 2030, but the solution recommended by the report as the most effective in enabling the EU to facilitate the best control over end-of-life ships was the use of Model B, whereby ships at risk calling at European ports or under European ownership could be more easily identified. In January 2011, the former port state control system was replaced by a New Inspection Regime, which is founded on a more risk-based approach to ship inspections, itself based on a variety of factors including previous inspection record, age of ship, and the performance of the ship operating company, the flag state and any organisation involved that works on behalf of the flag state.

A listing of vessels ready for dismantling is likely to be compiled from the various tools and databases such as Equasis, used by the European Maritime Safety Agency (EMSA)[20] for various purposes such as monitoring and vessel tracking systems, although it would be difficult to justify with certainty that such a listing would be 100 % accurate. However, the compilation of a list of ships likely to be ready for scrapping would appear to be fraught with practical difficulties considering the numerous and probably subjective factors that shape a decision to demolish, and legal implications on owners might lead to a move away from EU states flags.[21] Age does not necessarily determine the status of a ship as much as the question of

[20] EMSA is the authority that has responsibility for monitoring Port State Control at a European level in accordance with *Directive 2009/16/EC of the European Parliament and the Council of 23 April 2009 on port state control*. Other Directives of relevance include *Directive 94/57/EC on common rules and standards for ship inspection and survey organizations and for the relevant activities of maritime administrations* and *Directive 2002/59/EC of the European Parliament and of the Council of June 2002 establishing a Community vessel traffic monitoring and information system and repealing Council Directive 93/75/EEC*.

[21] ICS response to public consultation on the 2007 Green Paper. European Commission (2009).

maintenance over the ship's life and the number of old, but well maintained liners that went to the breakers after the last amendments to SOLAS became effective in May 2010 were the result of the cost of retro-fitting ships to the new requirements rather than the poor condition or age of the vessel.

Part B of the listings report considered the production of a list of shipbreaking facilities that could work up to and beyond those requirements defined principally in the guidelines of the *HKC* and the control and enforcement measures within the *Basel Convention* to produce what the report describes as a 'green and safe ship dismantling facility' (GSSDF). Additional factors had already been defined in a previous COWI study[22] as adding to the requirements that are not considered in the *HKC*, such as the treatment and disposal of hazardous wastes, third party audits, and workers' rights and international labour law conditions. Belgium's response to the public consultation on the EU's Green Paper was to question the extent to which the governments of the South Asian yards would monitor the compliance to the IMO standards of their recycling sites; the EU third party certification and audit scheme and the requirement of EU ships to go to certified facilities might offer guarantees that they are demolished in a manner that is environmentally sound.

A list of 24 suitable sites was contained within the report, 13 being located within the EU and 11 in China and Turkey. Sites listed as members of the International Ship Recycling Association (ISRA) (see Sect. 7.2.2 below) were considered to meet the criteria for inclusion in the listing. Although the report refers to the potential for great administrative cost of monitoring such listed sites, the principle has been carried forward to the new Regulation to enact the provisions of the *HKC* within European legislation that is currently passing through the EU adoption system. One proposal for addressing the issue of costs is via a tax abatement scheme related to either registrations or tonnage tax, although this is further complicated by the variations in the tax schemes that operate within the various Member States.

A major potential problem of inspecting ships under a regional scheme that may go beyond the requirements of the *HKC* is that of effectively creating a double standard for shipowners. Although the Convention does contain distinct provisions for Parties to take '*more stringent measures consistent with international law...in order to prevent, reduce or minimize any adverse effects on human health and the environment*,'[23] such a step on the one hand does not lead to a global level playing field; on the other hand, it is by allowing or even encouraging such distinctions that a process of continual improvement may be stimulated. Another potential problem may be imposed upon the question of shipbreaking capacity if shipbreaking states do actually rise to the spirit of the Convention and decline to authorize facilities deemed to be below defined standards.

[22] COWI, Lithauz (2008).

[23] *Hong Kong Convention*, Article 1.2 General obligations.

7.1.4 *The New Ship Recycling Regulation (SRR)*[24]

The question of equivalence between the two conventions was ultimately addressed by both the Parties to the *Basel Convention* and by the European Union and its Member States and in the end the suggestion made by the IMO (on the enactment of measures by the EU) became the path to be followed. In March 2012, a proposal for a Council Decision announced that by April 2010, the EU and its Member States had concluded that the *HKC*:

> ...appears to provide a level of control and enforcement at least equivalent to that provided by the *Basel Convention* for ships which are classified as waste under the *Basel Convention*.[25]

Moreover, the proposal concluded, and with apparent enthusiasm, that the new Convention:

> ...represented a major achievement for the international community, [and] provided a comprehensive system of control and enforcement from "cradle to grave" and strongly encouraged Member States to ratify the Convention as a matter of priority so as to facilitate its entry into force as early as possible and to generate a real and effective change on the ground.[26]

By October 2011, the Parties to the *Basel Convention* were also encouraging ratification of the *HKC*.[27] The question of equivalence is not simply restricted to an either/or situation; Annex Chapter 1, Regulation 3, requires that in implementing the requirements of the Annex, Parties shall take into account the relevant technical standards, recommendations and guidance as developed in the *Basel Convention* and by the ILO. This opened the way for new and formalised measures from the EU.

The proposed Council Decision recognised the ineffectiveness of current EU legislation to address problems of current shipbreaking practices, the high priority to the EU of environmentally sound management of ship dismantling and the fact that some provisions of the new Convention fall exclusively within the competence of the EU. It also recognised that there were no facilities within the OECD that could accommodate the demolition of the larger ships[28]—hence one of the reasons for the frequency with which European-flagged ships were often reflagged prior to demolition.[29] A new Regulation—initially referred to as the ad-hoc Regulation and subsequently as the *Ship Recycling Regulation*—was to be issued requiring

[24] Note the change in terminology to 'recycling' from the Commission's former preference for 'demolition'.

[25] European Commission (2012a), p. 2.

[26] European Commission (2012a), p. 3.

[27] European Commission (2012b), p. 4.

[28] That being said, it is worthy of note that the large ex-Royal Navy aircraft carriers HMS *Illustrious* and *Ark Royal* were broken up at Aliaga and the even larger ex-French carrier *Le Clemenceau* at Able UK, Hartlepool.

[29] Blanco (2012).

Member States to ratify the Convention without delay,[30] early ratification or accession by the Member States thereby adding an incentive to other non-EU states for ratification of the Convention and its early introduction into force.[31] The choice of Regulation over Directive meant that the instrument could become effective within all Member States without the delay of enacting the measures into national legislations. The time scale of effectiveness is therefore much shorter than could be expected of the IMO Convention—indeed some IMO Conventions, although adopted, have not entered into force at all.

The *SRR* is intended to facilitate an early introduction of the *HKC* by enacting the principles of the Convention's basic provisions into EU law, recognising that the existing 2006 Regulation on shipments of waste[32] (and thereby the *Basel Convention*, including the BAN amendment, upon which it is based), has been '*almost systematically circumvented by EU-flagged ships.*'[33] The 2006 Regulation, in as much as it applies to ships, will be replaced by the new *SRR*.

Within the definitions contained in the new Regulation is the definition of 'shipowner,'[34] which includes '*the natural or legal person owning the ship* ***for a limited period*** [emphasis added][35] *pending its sale or handover to a ship recycling facility*,' a definition obviously aimed at cash buyers, whose liabilities as owners have often been disputed in the past. Further, the definition of 'ship recycling'[36] covers '*the complete or partial dismantling of a ship...in order to recover components and materials for reprocessing*' and includes the storage and treatment etc., of those materials, but excludes '*their further processing or disposal in separate facilities*,' *i.e.* the recycling definition specifically excludes the actual recycling activity (at the neighbouring rolling mills etc.), which may not appear to sit easy with European case law on recycling.

The new Regulation is to apply to the whole life cycle of EU-flagged ships, which would only be allowed to be recycled at approved and listed facilities—reference the 2010 listing report above.[37] The Regulation is to include even more stringent environmental requirements than required by the *HKC*, prior to the entry

[30] The latest time envisaged for ratification is 2020, at which time the *HKC* was expected to come into force. European Commission (2012c), p. 32.

[31] European Commission (2012c), p. 28.

[32] *Regulation (EC) 1013/2006 of the European Parliament and council of June 2006 on shipments of waste*.

[33] European Commission (2012b), p. 2.

[34] *Proposal for a Regulation on ship recycling*, Article 2 (13) Definitions.

[35] Note that the new Bangladesh *Shipbreaking and Ship Recycling Rules* also makes special reference to the position of cash buyers who exercise limited time ownership—there now appears to be a general intention that full liability of cash buyers is formally recognised.

[36] *Proposal for a Regulation on ship recycling*, Article 2 (5) Definitions.

[37] This Regulation on end-of-life ships does not equate to other European mechanisms for end-of-life vehicles, which places responsibility for disposal with the producer. The latter is regional producer law, but shipping is international and no producer responsibility is associated. Comments from discussions with recycling facilitator 2.5.2012.

into force of the Convention.[38] These include specific requirements for the recycling facilities; the creation of a European list of (approved) ship recycling facilities; and the establishment of a contract between the facility and the ship-owner. The Regulation also seeks to address one of the barriers to observing the 2006 *WSR*, *i.e.* the very limited capacity for shipbreaking within the OECD, by allowing ships to be demolished in facilities that are outside the EU but which meet the specifications to be defined and are included in the European list.[39] The scope of ships covered by the instrument, requirements for IHMs and surveys and the need for a specific ship recycling plan, are basically as per the *HKC*. Before recycling can take place, ships for disposal are to be in possession of a recycling certificate issued by their Member (flag) State; tankers must also have cargo tanks and pump rooms in a condition that is ready for certification as safe-for-entry and safe-for-hot-work. These conditions are the responsibility of the shipowner.

Included in the requirements for the facilities, which are to be designed, constructed and operated in an environmentally sound manner, is the need to '*demonstrate the control of any leakage, in particular in intertidal zones*' (this requirement, by implication, recognising the continuation of the beaching system although specific reference to beaching has been studiously avoided in the Regulation), and '*handle hazardous materials and waste only on impermeable floors with effective drainage systems.*'[40] A further requirement that operations are conducted from built structures was subsequently added by the European Parliament, although the definition of a built structure was not included.[41] By stipulating further that waste facilities that receive wastes from the recyclers are themselves authorised to treat and dispose of them in a manner that does not endanger human health or the environment, the Regulation notes that the standards of the waste facilities are to match those in force in the EU. A requirement to ensure that all hazardous wastes generated are traceable was vetoed by some of the major shipping states such as Malta and Greece. Requests for inclusion on the European list are to be voluntary and not arranged via the state as per the Convention; all those requesting inclusion thereby also accept inspection by the EC or its agents on or before inclusion on the list.[42]

---

[38] European legislation that goes beyond the demands of international treaties and prompts elaboration is not new; the EU measures with regard to the phase-out of single-hulled tankers prompted the IMO to amend *MARPOL* 73/78 and apply the more stringent EU conditions to all tankers worldwide, and the proposed EU action on compensation for oil pollution damage led the IMO to establish the International Oil Pollution Compensation Supplementary Fund.

[39] Until the European list is published, European-flagged ships are only to be recycled in facilities located in the EU or in an OECD state. Thereafter, ships may be recycled at any facility included in the list.

[40] *Proposal for a Regulation on ship recycling*, Article 12 (l) and (m), *Requirements for ship recycling facilities*.

[41] It was argued, for example, at the 8th Annual Ship Recycling Conference in September 2013 that carrying out work within the hull of a vessel being demolished might actually constitute working within a built structure.

[42] Article 15.8, Ship recycling facilities located outside of the Union.

To date, the second way in which the effectiveness of current legislation has been circumvented has been the ease with which shipowners have been able to sell their end-of-life vessels to third parties (usually cash buyers), who may then immediately sell the ships for scrapping. Provisions to counter this route—as well as other failures to meet stated provisions of the Regulation—have been incorporated in the form of '*effective, proportionate and dissuasive penalties*' that Member States will be obliged to enforce. The owners of ships sent to recycling facilities that are not included in the European list are to be subject to a penalty which, as a minimum, corresponds to the price paid to the shipowner for its ship.[43] Ships which are sold, and within less than 6 months are sent to facilities not included on the European list, shall result in penalties imposed *jointly* on both the last and the penultimate owner if the ship is still flagged to a Member State or imposed *solely* on the penultimate owner if the ship is flagged to a state other than a Member State.[44] Repeated sales of an end-of-life vessel within a short period of time may be regarded as avoidance. Member States are to notify the Commission of their national legislation that relates to enforcement of the Regulation and the penalties applicable.[45] However, a meeting of the European Parliament's ENVI Committee[46] on 26 March 2013 voted to impose penalties on the sale of ships broken up by beaching or scrapped at non-EU approved yards within a period of 12 months after sale in an attempt to block reflagging as a means of avoiding the new requirements.

The above measures are intended to act as a stimulant to persuade owners to make early use of facilities that are deemed to be safer and more environmentally sound at shipbreaking than others. The only effective influence that the EU (and others) may have over the procedures and conditions of the breakers is the influence that they have over their clients, *i.e.* the shipowners,[47] yet neither the *HKC* nor the proposed *SRR* define just when a ship may be considered ready for scrapping, although the Regulation acknowledges the number of different factors that may instigate such a decision. The Regulation also acknowledges the difficulties in identifying such decisions and applying the requirements of the *WSR* if a ship happens to be in international waters or in the waters of the recycling state. Although the attraction of environmentally safer shipbreaking has been increasing with the growing capacity for the same, this is muted by the price differential resulting from higher overheads of these yards.[48]

A further reason behind the EC's move to introduce the *SRR* is that ratification of the *HKC* is open only to states and not to the EC; the *SRR* is therefore intended to obviate any problems that may arise from Member States ratifying the *HKC* at

[43] Article 23.2 Enforcement in Member States.

[44] Article 23.5 Enforcement in Member States.

[45] Article 23.7 Enforcement in Member States.

[46] The European Parliament's Committee for Environment, Public Health and Food Safety.

[47] Comment from discussion with IMO representative, 1.12.2009.

[48] McCarthy (2012a), p. 3.

different times, thereby for a period establishing different regimes and problems of competition within the EU.

Included within the *HKC* is the option for Member States to incorporate national measures which extend those of the Convention itself. Whilst such a provision may be defended in the name of ongoing improvement, it can also create uncertainty in as much as it can engender dual standards, such as the restriction on extra substances like as Perfluroctane Sulphuric Acid (PFOS) and its derivatives.[49] It is considered that any enhancements of the Regulation over those contained in the *HKC* could only delay rather than assist the eventual ratification of the latter.[50]

However, there is little within the scope of proposals to ameliorate the impacts of current shipbreaking practices that does not generate a strenuous response from some interested party and such is the case with the proposed port tax on all visiting vessels. Strong opposition was voiced by the European Sea Ports Association; the European Community Shipowners' Association; the Asian Shipowners Association and the International Chamber of Shipping, who collectively consider that such a proposal will work as a barrier to the introduction of the *HKC*; offend Europe's trading partners; jeopardise the jobs of many in the developing countries and promote an incentive towards the transfer of freight to road and rail and away from sea-borne transportation.[51] Not stated, but presumably within those arguments, is also the impact that such a levy may have upon the shipowners' operating costs. Whilst the Commission's proposed levy had been well supported by the Environment Committee of the European Parliament, the measure was rejected at its first reading by the European Parliament on 18 April 2013. The timeline for the preparation of new proposals for a financial mechanism was reset at mid 2014/2015.

In addition to the above two reports on a funding mechanism, a third study was issued by the Dutch economic research consultancy Profundo in January 2013.[52] Commissioned by the NGO Shipbreaking Platform, the study recognised the threat to the proposed S*RR* from reflagging away from Member States as well as the need for a financial mechanism to support the Polluter Pays Principle and considered three alternatives that the scheme might follow, all being based on ship visits to European ports rather than on the basis of flag or ownership. Option 1 is the Ship Recycling Fund, built up from fees levied on ships visiting European ports and payable to owners of European ships in the form of the different (additional) cost of recycling at an approved facility. Option 2 is a Ship Recycling Insurance scheme established with an accredited insurance company. Premiums paid to this insurance fund would cover the additional cost incurred with recycling at an approved facility, paid out to shipowners on proof of such demolition. Option 3 is a Ship Recycling

[49] This is in line with the Commission *Regulation (EU) 757/2010 of 24 August 2010* amending *Regulation (EC) No 850/2004 of the European Parliament and of the Council on persistent organic pollutants* as regards Annexes I and III.

[50] Hintzsche (2013).

[51] Eason (2013), p. 1.

[52] Profundo (2013).

Account together with a Transitional Fund under which a shipowner makes an annual deposit to the account, receiving an annual certificate which allows the ship to enter European ports; the accumulated money is paid back to the owner on proof of responsible recycling. A temporary surcharge on all the deposits is to be transferred to a Transitional Fund to subsidize the responsible recycling of older ships. Profundo recommends Option 3 as having fewer disadvantages than the others, although it acknowledges that further research is required on the details of the scheme.

However, the validity of the Regulation has been challenged by two legal opinions, issued by the former chief counsel to the European Commission, Dr Ludwig Krämer,[53] and by the Centre for International Environmental Law (CIEL).[54] In brief, these opinions state that both the *Basel Convention* and the *Basel Ban* have been incorporated into European legislation via *Regulation 1013/2006*, which is held to include ships destined for scrap.[55] By seeking to remove ships from the scope of the latter, the EU unilaterally is seeking derogation from *Basel*, which includes no provisions for derogation, even of a single provision, nor does the Convention accept revisions.[56] The scope for possible separate bilateral, multilateral or regional agreements further requires that any such agreements shall not permit conditions which are less contained in the Convention or derogate from the environmentally sound management that it prescribes.[57] The proposed *SRR* would become secondary legislation, and since EU Treaty law as primary law ranks higher than international agreements which, in turn rank higher than secondary law, (namely Directives, Regulations and Decisions), the *SRR* could not cancel or amend the regime established by *Basel*. The legal opinions also define that the *SRR* does not have equivalence to *Basel* since it is remiss in providing for PIC, for the criminalization of illegal transfers of hazardous waste and the obligation on exporting states to re-import illegal shipments. Further, the *HKC* excludes certain categories of vessels and fails to provide control over waste downstream of the breaking facilities.[58] By standing in opposition to the international agreement represented by *Basel* and the *Basel Ban*, the proposed *SRR*, as secondary legislation, is deemed to be illegal under international law and under EU law and, under Article 263 of the Treaty on the Functioning of the European Union, is also deemed to be a candidate for annulment by the ECJ.[59]

In addition to the two independent legal opinions above, the Legal Service of the European Council published its own opinion in November 2012,[60] stating that since

---

[53] Krämer (2012).

[54] Centre for International Environmental Law (2012).

[55] This has been held to be the case by the judgements of the national courts as described in Chap. 5.

[56] Centre for International Environmental Law (2012), p. 7.

[57] *Basel Convention* Article 11.

[58] Centre for International Environmental Law (2012), p. 8.

[59] Centre for International Environmental Law (2012), p. 11.

[60] Council of the European Union (2012).

the Basel COP 10 encouraged Parties to ratify the *HKC*, this justified the view that the '"*preliminary assessment*" of the EU and its Member States amounts to a good faith interpretation of the *Basel Convention*.'[61] It also noted that there is capacity for recycling EU-flagged vessels in the EU that is currently not used since sales to breakers elsewhere result in higher returns for ship owners, a factor which makes establishing a business case in the EU very difficult—this consideration failing to meet with the requirements of Article 11 of *Basel* since it is essentially of an economic rather than environment consideration.[62] there is a serious risk that excluding ships from *Regulation 1013/2006* as proposed could amount to a breach of the obligation not to defeat the object and purpose of the [Basel] treaty.

Despite these arguments, the EU appears determined to proceed with the introduction of the *SRR*. Guidelines to the Regulation are to be published in early 2014 and the feasibility study relating to the financial mechanism by 2014/2015. The Regulation itself is expected by late 2013/early 2014 and will come into force 20 days after publication in the Official Journal. The list of approved sites is also to be published in the Official Journal, some 36 months (at the latest) after the Regulation has come into force, *i.e.* by the end of 2016. The *SRR* is intended as a bridging measure between the *WSR* and the *HKC* until the latter comes into force; thereafter, the *SRR* will need revisiting.

7.2 Other Commercial Initiatives

In addition to the EU's impetus to promote and mobilise environmentally safe shipbreaking, there have been strong moves by independent, commercial operations to promulgate similar approaches, which they see as either an ethical way forward, or a growing business opportunity, or perhaps both, and a number of these are examined below, although other enterprises are in effect. As with other control and regulation instruments, the adoption and eventual coming into force of a new convention sets off a number of resultant or supporting measures and activities; consider, for example, how producer responsibility measures over packaging waste basically changed a seller's market in waste paper to a buyer's market and a hitherto uncomplaining car breaking industry suddenly becoming more circumspect about its actual margins once the motor manufacturers were identified as ultimate cost insurers, following end-of-life vehicle regulations. With the arrival of international provisions for shipbreaking, on the other hand, new opportunities were seen in both facilitating measures to help the various actors adapt to the new regime as well as offering a more long-term opportunity for enhanced business portfolios. The emergence of facilitated shipbreaking operations not only aids the shipowner in

[61] Council of the European Union (2012), para 21. Here the Opinion refers to accord with Article 31 of the Vienna Convention on the Law of Treaties.

[62] Council of the European Union (2012), para 27.

navigating what will become an increasingly important selection of breakers, but also addresses directly the need to demonstrate the now-demanded corporate social responsibility. This facilitated aspect of shipbreaking has spread from an individual company's desire to demonstrate its own credentials to an independent sub-industry that, although still small, may prove to be a model for growth.

Whether the catalyst for some of these developments was the demolition of the *Sandrien* in dry dock in Amsterdam, as suggested by an interviewee, or whether a specific individual has been a common factor in more than one organisation, the fact remains that a number of these enterprises have originated or been based in the Netherlands.

The examples listed below are not intended as an exhaustive list, but are included to demonstrate the nature of some of the initiatives currently being developed.

7.2.1 Maersk/Sea2Cradle/Wilhelmsen

The pioneering approach by P&O Nedlloyd—later to be enveloped into AP Møller-Maersk—to establish firm links with China increasingly is becoming a model of ethical shipbreaking, although currently owners opting for the more environmentally sound yards are very much in the minority. Since 1997, by establishing exclusive arrangements with nominated shipbreaking yards to demolish that company's ships, providing technical advice and PPE for the workers and supervision of the process, and despite receiving payments for their end-of-life vessels which are below those that might have been expected from breakers in the sub-continent, Maersk has been able to ensure that their ships are disposed of in a manner that has lessened the impacts on both the workers and the surrounding ecology. Although shipbreaking in China may not yet be at the stage that might be considered environmentally sound, such arrangements may be said to have moved shipbreaking (at least in China) closer towards the concept of sustainability. Although not required until 2009, Maersk has been compiling IHMs for its ships since 2007.

The success of this approach led to the formation of a department in the company, which was subsequently able to offer a similar service to others shipowners who sought to introduce an element of corporate social responsibility into their operations. Maersk Ship Management Recycling was established in 2000 as an independent entity within the Maersk group. As well as controlling the disposal of its own end-of-life ships, Maersk also undertook the disposal of ships for other shipowners and by 2010 had already recycled over 50 ships in this way, half of them belonging to other owners.[63] With the increase in the number of ships, especially

[63] For this achievement, Maersk was awarded the Environmental Protection award at Lloyd's List Asia Awards in 2010. Moller-Maersk Group nd. Although Maersk operates a large fleet of (mainly) container ships, usually only three or so are sent directly to the breakers each year; many ships are sold to other operators before they reach their retirement age.

bulkers, tankers and eventually cruise ships, expected to reach the breakers, due partly to the increase in the number of new builds and partly to the need to dispose of single-hulled tankers, Maersk initially saw this service to other shipowners as not only a profitable business opportunity, but also something that was necessary to maintain its own level of expertise.[64] The service consisted of assistance with the provision of an IHM and an appropriate ship recycling plan with the chosen breaker, documentation for the delivery of the ship, including a sales contract, and supervision during the actual demolition to ensure that this was consistent with the plan formulated.

Despite its success, in April 2011 Maersk announced that it would withdraw from such activities as this was no longer deemed to be core business. Instead, these activities would be performed on behalf of Maersk and others by the newly-formed organisation known as Sea2Cradle, an organisation headed by a key figure in the former Maersk Ship Management Recycling operation. In addition to using at least four yards in China, Sea2Cradle also intends to use a shipbreaking facility in Ghent, which has disposed of various ex-Royal Navy vessels in recent years, and possibly yards in Turkey. As well as providing assistance with documentation for the arrival of vessels and after demolition for the previous shipowner, Sea2Cradle will also ensure proper pre-cleaning prior to demolition, which takes place at the breaker's site, the company having contracts with both the former owner and the shipbreaking yard that allow it to call a halt to demolition, should it be decided that appropriate standards were not being observed. Sea2Cradle will also be involved with the provision of training at the yards.[65] In this manner, Sea2Cradle, as well as Wilhelmsen and Green Recycling Services (see below), act as agents for the former shipowners. As ships are being scrapped, Sea2Cradle compiles a photographic record of the stripping and demolition process which, with the ship recycling plan and copies of the documentation transferring hazardous waste from the breaker's site to the waste handling facility, comprise proof of the process for the former shipowner.[66] The 70+ ships listed as demolished via the process since 2001 include those demolished by the same team when they were a part of P&O Nedlloyd, and subsequently Maersk.

Amongst others recognising this new opportunity is the Wilhelmsen Group, which operates its own large fleet of vessels, particularly car carriers. Wilhelmsen Ship Management provides a similar service for '*socially responsible Green ship owners*' with closely supervised yards in China which can operate to the standards that are encapsulated within the new Convention and where are based Wilhelmsen representatives, who can halt the demolition process if it appears to be deviating

[64] Garfield (2010).

[65] Safety4Sea (2011).

[66] Although the transfer notes do not actually provide evidence of the final and environmentally safe disposal of the waste, it is as close as one might obtain without actually imposing upon Chinese national sovereignty. Comments arising in discussions with Sea2Cradle representative 2.5.2012.

from agreed standards. Demolition takes place against a pre-defined set of Key Performance Indexes (KPI), which in turn will determine the payment of an agreed bonus incentive payment to the yard.[67] As well as the selection and vetting of yards, their Green Ship Recycling process also involves ensuring that vessels are pre-cleaned prior to their arrival at the breakers, with the removal of all hazardous materials and wastes. Pre-cleaning in turn requires the availability of an IHM, which is another service that Wilhelmsen offers to owners. Traditionally, shipowners have disposed of end-of-life vessels to cash buyers—usually via a broker (unless they have a dedicated sales staff for this purpose). The cash buyer then starts negotiations with breakers for the optimum price for the ship. Using a 'green' ship recycling specialist creates a more direct link with the breaker committed to operating against more closely defined standards.

In addition, numerous other organisations including classification societies and marine consultants of various kinds are also extending their portfolios to the preparation of IHMs and surveys to validate the same. This relatively new method of selling and scrapping ships via the services of a professional and dedicated management company is one that is likely to expand, given the tightening of regulatory controls, and the growing perceived need for owners to dispose of their unwanted and end-of-life ships in a manner that demonstrates some degree of corporate social responsibility, and represents a new, third way that might be termed '**facilitated disposal**.' Whether this expansion is prompted or seen as an ethical move and a step towards greater social responsibility, or merely as a fortuitous business opportunity, is perhaps a secondary issue; this development of a more responsible approach to shipbreaking is a matter that can both help to secure the coming into force of the Convention or something that can work either in parallel or even instead of it; either way, it should prompt enhancements in the standards to which yards are currently working.

Another scheme involving a Danish shipowner with a close association with a Chinese yard was demonstrated by the sale for demolition of the DFDS's[68] ro–ro vessel *Tor Anglia*, clauses in the sale contract reserving the right of DFDS to approve the yard and requiring that the contaminated bilge water that was in the vessel on its arrival was to be shipped back to Scandinavia in containers. This is an example of waste control that could enter into more demolition contracts as regulations, especially European regulations, become more widespread and enforceable; it also demonstrates recognition by the company of its liabilities. In this particular instance, however, it was stated in interview that the shipbreaker had adequate facilities on site to cope with safe disposal and such an arrangement was somewhat unnecessary but, where the shipbreakers' facilities are more limited, certain repatriation of hazardous wastes might offer an acceptable course.

[67] Wilhelmsen nd.

[68] DSDF is an abbreviation of Det Forenede Dampskibs-Selskab (The United Steamship Company), a major Danish ferry company.

7.2.2 International Ship Recycling Association (ISRA)

Formed almost 2 years before the adoption of the *HKC* but following parallel principles with regard to the breaking sites, the ISRA seeks to unite yards from around the world that subscribe to ISRA developed standards[69] aimed at sustainability. Successful members are able to market their services to those 'responsible' shipowners under a brand programme that is part of a unified promotion whereby ISRA actively encourages shipowners to follow a more socially responsible path of end-of-life ship disposal. The aims of the Association go beyond merely representing a group of shipbreakers, and include demonstrating methods that differ from those traditionally practised in South Asia and promoting an improved image, promoting alliances with shipowners and presenting a common voice at the IMO. In addition to meeting (at least) the Association's standards, members are also required to observe a further raft of measures that include requirements for the exchange of information and technical co-operation, reporting, the use of ship recycling plans etc.

Having a growing membership that is distributed worldwide[70] reflects the Association's announced intention to address both the industry and legislation—especially that which currently allows beaching—at an international level, the latter issue being a topic that shipbreakers have not, to date, entered into directly. The ISRA has an office within the Sea2Cradle offices, showing the close link between the two and the leading role played by the owners of Sea2Cradle over recent years in 'facilitated disposal.'

7.2.3 Green Recycling Initiative

Approaching the shipbreaking business from the customer end of the industry, and looking beyond the simple shipowner/shipbreaker link, is the Green Recycling Initiative (GRI), which in 2009 announced the formulation of a US$300 million fund to purchase some 120 end-of-life dry bulkers, with the aim of enabling steel producers to source traceable scrap whilst providing a more environmentally sound disposal process for the owners of ships destined for scrap.[71]

---

[69] Amongst the minimal standards compiled by the Association are requirements for the A level members to have valid ship recycling permits issued by the appropriate state authority; adhere to the guidelines as issued by the IMO, the ILO and the Basel Convention, which must be covered by laws of the state that are compliant to those guidelines; provide suitable barriers to prevent all aspects of the environment from the release of pollution; adequate control for the collection and disposal of hazardous materials and wastes; the provision of proper PPE, training, accommodation and emergency services for the workers. International Ship Recycling Association (2008).

[70] The ISRA has members located in China, Turkey, USA and the Netherlands.

[71] The Maritime Executive (2011).

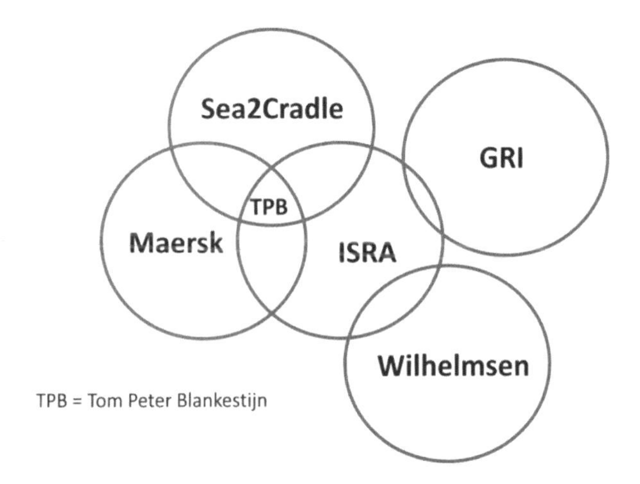

Fig. 7.1 Organizational links in ship disposal. Source: Galley (2013)

Based in Singapore, GRI is seeking investors from both steel makers and shipowners, and aims to construct its own dock-based shipbreaking facilities in areas that can safely accommodate vessels awaiting demolition; in the interim, the company plans to use such shipbreakers as are members of ISRA.[72]

Although the scope of the new ventures outlined above may vary in extent, they all share common ground in trying to promote a more responsible approach to shipbreaking and a closer integration between the various industries involved. Further, there are distinct organizational links between them (see Fig. 7.1), with three of the organisations having a common factor in the person of Tom Peter Blankestijn, originally of P&O Nedlloyd and subsequently Maersk, and co-founder of Sea2Cradle, which together with Wilhelmsen are associates of the ISRA.

7.2.4 GreenDock

Distinct from the above organisations, and one that is directly engaged in shipbreaking, is GreenDock. The GreenDock process[73] is essentially a 'package,' whereby ships are demolished in dry dock or on a floating pontoon. In this manner, GreenDock claims that ships can be demolished within a period of 20 days, at the same time containing any waste discharges that are generated. The process not only covers the provision of physical facilities, but also incorporates a management

[72] The Maritime Executive (2011).

[73] A schematic of the project on the company's website shows a ship being winched stern first up the sloping floor of a dry dock. Dock gates are then closed behind the bows of the ship and the floor is raised until level. The ship is then progressively cut into tranches, each one being carried out of the dock on a wheeled vehicle to a row of cranes behind the dock where the tranches are further demolished. Patent requested. GreenDock nd.

system and a data collection system that will enable both the former ship owner and the future customer for the scrap materials (the rolling mills) to track the demolition of the vessel. This method effectively replaces beaching with demolition in a confined environment, but thereby incurs a heavy capital investment. Shipbreakers are to be inspected and certified by the independent body of KEMA[74] and the demolition process is subject to monitoring both before and during the actual demolition. In 2010, a US$51 million project was proposed for Bangladesh, with other potential projects in Vietnam, Cambodia, India, Thailand, UAE and Curaçao being also mentioned, these states perhaps representing likely new locations for emergent shipbreaking industries.[75]

Given their choice of ship demolition on hard surfaces, GreenDock would obviously prefer to see the abolition of beaching included within the Convention and similarly champions the extension of the scope of the Convention to include ships of less than 500GT, which it believes could be demolished in the many small EU yards that could meet ISO standards.[76] Like Maersk, Sea2Cradle and ISRA, the organisation is located in the Netherlands.

7.2.5 China

China is taking strong and positive steps to build upon its growing reputation as a model of shipbreaking practice with a project that offers perhaps the first purpose-built, large and modern facility. At Changxing Island, in the northern province of Liaoning, Dalian, China's largest state-owned shipbuilder has constructed the Dalian Shipbuilding Industry Ship Recycling yard, an investment of US$500 million, providing a 460,000 m^2 facility with some 1,650 m of recycling quay that will have the capacity to dismantle 100 ships of 20,000–300,000 DWT per year, several being demolished simultaneously. The process is carried out in three stages; the upper parts of ships are removed in sections by crane whilst the vessel is moored alongside, the remainder of the hull then being demolished in dry dock, whilst subsections are finally processed on land. In this manner, the facility, in common with a number of other Chinese yards, will carry out a demolition process that reflects the shipbuilding process in reverse, rather than the simple slicing up of a ship that is more typical of the beaching method. Facilities are provided for all wastes to be handled on site, with special provisions being made for sewage, asbestos and incineration. The production of large volumes of scrap will ease the

[74] KEMA is a global energy consultancy based in Arnhem, Netherlands offering consultancy, operational support, testing and certification services.

[75] Integrated Regional Networks (2010) and GreenDock nd. IRIN is a non-profit organ of the UN Office for the Co-ordination of Humanitarian Affairs (OCHA), and is based in Nairobi.

[76] GreenDock response to the public consultation to the 2007 *Green Paper*. European Commission (2009).

demand for imports of iron ore into China, a trade that has been running at high levels in the recent past.

Dalian's new facility represents a good example of vertical integration of similar activities in the ship handling world; as well as shipbreaking, shipbuilding and ship repair work are also to be undertaken on site, which will be able to switch resources between activities as market demands change. Some of the scrap generated will be melted for re-use in Dalian's shipbuilding activities, which are also on site, effectively providing what they described as a 'one-stop shop' for vessels, from building, through repair, to ultimate scrapping. Shipbreaking at Changxing will be at the high end of the market, but the yard should benefit from the fact that it can call upon a captive market represented by obsolete Chinese ships. By taking an 18 % share in the yard,[77] the Pacific International Lines (PIL) container line of Singapore will be guaranteed space for the repair of its 105 ship fleet.[78] PIL is already a customer of Dalian's shipbuilding activities, and this further consolidation is likely to lead to a strong connection with the new yard for eventual demolition of PIL ships. Trials started in November 2012 and production began the following year. The facility is recognised within the EC's Proposal for Regulation as:

> ...the result of actions taken by the public authorities to promote specifically the green ship recycling market and close substandard facilities as well as investments made by responsible European shipowners in safe and sound recycling facilities.[79]

This flexibility of switching between activities would be in accord with the call by some within the shipping industry for China to switch some of its extensive shipbuilding capacity to shipbreaking. Chinese shipbuilding yards expanded rapidly during the earlier boom years of shipping and there are some expectations that China will endeavour to maintain full employment for its builders, working in opposition to the pressures to reduce fleet sizes that is pervasive amongst shipowners today.[80] Another report cited more than half of China's shipbuilding yards are likely to go out of business '*in the next two or three years*.'[81]

In March 2013, the Irish-owned CEMS-Dalian Industry Company Ltd. also announced its intention to develop ship dismantling and recycling facilities in Zhuanghe, near Dalian, where wastes will be recycled or disposed of '*to the highest standards of environmental safety*.'[82]

---

[77] A further 67 % of the yard is owned by Dalian Shipbuilding, and the remaining 15 % owned by Angang Steel.

[78] China Shipping (Group) Company (2012).

[79] European Commission (2012b), p. 3.

[80] McCarthy (2012b), p. 3.

[81] Shipping World and Shipbuilder (2012), p. 8.

[82] Cork Chamber (2013).

7.2.6 Japan

Japan in the past has attempted various forays into establishing shipbreaking operations in other countries on the basis of the scrap steel being returned to Japan whilst the waste products remain in the host countries. A new pilot project, however, was undertaken in Japan in 2010 to establish whether shipbreaking could re-enter Japan after an absence of some two decades. At the port of Muroran in Hokkaido, a Japanese registered K Line car carrier *New York Highway* arrived to be demolished using modern technologies to determine whether safe and environmentally sound processes are viable. The chosen breaker was Teraoka Co., a member of the Muroran Ship Recycling Study Group and the project was subsidized to the sum of US$3.37 million by the Japanese government. The trial was organised by Japan's Maritime Bureau of the Ministry of Land, Infrastructure, Transportation and Tourism to test new methods of ship recycling. It was reported that if the project proves to be feasible, it will act as a test project for the development of larger shipbreaking facilities and that the technical information obtained would be shared with other shipbreaking industries, but it was not clear whether this project was devised to re-foster shipbreaking in Japan or as an incentive elsewhere.[83] After the completion of the exercise, which took some 9 months, Japanese representatives were invited by an interviewee promoting facilitated shipbreaking to see how the activity was being completed in 8 weeks in China.

Following a February 2012 meeting between Japan's Ship-owners Association and the Chief Minister of Gujarat to discuss a project aimed at raising the facilities across the 173 yards at Alang, an Official Development Assistance loan of US $230 million was made to the Gujarat Maritime Board by the Japan International Co-operation Agency. This loan, together with appropriate technology from Japan, is aimed at raising operations at Alang to the standards required by the Convention[84]; amongst the provisions referred to are dry docks to allow the removal of hazardous wastes and hard surfaces to allow large blocks to be lifted. Currently, Japan is reliant on China to meet its shipbreaking needs.

7.2.7 Others

The public consultation on the 2007 *Green Paper* elicited a number of responses from Member States, NGOs, Trades Unions, private enterprises and individuals. Whilst the majority of these responses related to the measures proposed by the EC, and have been referred to above and in the preceding chapter, some suggestions related to specific initiatives that might be physically introduced at the shipbreaking sites. Amongst these, the suggestion from Bertech[85] was for the creation of a fleet of

---

[83] Leach (2010).

[84] Fairplay (2012), p. 29.

[85] European Commission (2009).

floating Ship Recycling Support Vessels (SRSV), either converted or purpose built vessels moored off the shipbreaking beaches, which do not have the requisite waste handling facilities, to accommodate the removal and handling of hazardous materials from ships, cleaning of tanks for hot work certification etc., just prior to demolition. Bertech also champions the adaption of telescopic equipment currently employed in cleaning and painting ships under repair, using plasma torches for cutting up ships either in dry dock or in water.

There is a wide range of initiatives that are under way in various arenas to promote or enhance the aims and objectives of the *HKC*. Whilst the Convention may have acted as a catalyst to such developments, they in turn may promote and facilitate the coming into force of the Convention—or may merely continue to develop as new and independent options in ship recycling. Either way, the adoption of the treaty has demonstrated recognition of the liabilities surrounding end-of-life ships, a new international call for improvements, and thereby an international market for new options in advancing the state of shipbreaking.

References

Blanco S (2012) EU regulatory review. In: 7th annual ship recycling conference, June 2012, London

Centre for International Environmental Law (2012) Legality of the EU Commission proposal on ship recycling. December 2012, Geneva

China Shipping (Group) Company (2012) PIL moves into yard game. www.ssoa.org.cn/en/home/shippingnews/24715.shtml. Accessed 4 Apr 2012

Cork Chamber (2013) CEMS-D announces plans to develop ship dismantling and recycling facility. www.corkchamber.ie/news/2013/03/26/CEMSDALIAN-ANNOUNCES-PLANS-TO-DEVELOPDISMANTLING-AND-RECYCLING-FACILITY. Accessed 20 Mar 2013

Council of the European Union (2012) Opinion of the Legal Service 16995/12 28 November 2012

COWI, Lithauz (2008) Study on the certification of ship recycling facilities. Final report. EMSA, Lisbon

Eason C (2013) Shipping rallies against Brussels ship-recycling tax. Lloyd's List, 19 April 2013, London

European Commission (2007) Green paper on better ship dismantling. COM(2007) 269 final

European Commission (2009) Details of EC (and other) documentation on ship dismantling. http://ec.europa.eu/environment/waste/ships/index.htm. Accessed 20 Nov 2009

European Commission (2010) An assessment of the link between the IMO Hong Kong Convention for the Safe and Environmentally Sound Recycling of Ships, the Basel Convention and the EU Waste Shipment Regulation. Communication from the European Commission to the Council COM(2010)88 final

European Commission (2012a) Proposal for a Council decision requiring Member States to ratify or to accede to the Hong Kong International Convention for the Safe and Environmentally Sound Recycling of Ships 2009, in the interests of the European Union. (Explanatory memorandum) COM(2012) 120 Final

European Commission (2012b) Proposal for a Regulation of the European Parliament and of the Council on ship recycling. Explanatory memorandum COM(2012) 118 Final

European Council (2012c) Commission staff working document. Impact assessment accompanying the document Proposal for a Regulation of the European Parliament and of the Council on ship recycling. SWD(2012) 47 Final. 23 February 2012, Brussels

European Maritime Safety Agency Ship Recycling Workshop (2011) 27–28 June 2011, Lisbon
European Parliament (2009) European Parliament resolution of 26 March 2009 on an EU strategy for better ship dismantling. Preliminary
Fairplay (2012) Japan offers to raise Alang standards. Fairplay, 1 March 2012, London
Galley M (2013) Hazardous materials in shipbreaking – where do the liabilities lie? PhD research thesis, Southampton Solent University, Southampton
Garfield G (2010) Maersk Bets on its China recycling expertise. www.gmsinc.net/gms/news/tradewinds/Maersk%20bets%20on%20its%20china%20recycling. Accessed 30 Mar 2012
GreenDock (nd) GreenDock. www.greendock.nl/files/greendockfolder.pdf. Accessed 10 Apr 2012
Hintzsche W (2013) Shipowner's perspective. Responsible recycling and the need for a level regulatory playing field. In: 8th annual ship recycling conference, 26–27 September 2013, London
Integrated Regional Networks (2010) Bangladesh: taking toxins out of ship-breaking. www.irinnews.org/Report/90376/BANGLADESH-Taking-toxins-out-of-ship-breaking. Accessed 15 April 2012
International Ship Recycling Association (2008) ISRA Standards for A members, including Annex 1 to Standards – Additional requirements for ISRA members. Adopted by the 2nd General Meeting of the ISRA, London March 31st 2008. www.isranetwork.com/mn_uploads/ISRA_additional_requirements_April_16th_2008.pdf. Accessed 1 Dec 2011
Krämer L (2012) The Commission proposal for a Regulation on ship recycling, the Basel Convention and the protection of the environment. 4 November 2012. www.shipbreakingplatform.org/shipbrea_wp2011/wp-content/uploads/2012/11/Ludwig-Kraemer-Legal-Analysis_EC-Proposal-ship-recycling-regulatio-Nov2012.pdf. Accessed 10 Sept 2013
Leach PT (2010) "K" Line to test ship-scrapping system. J Commerce [Online 9 March 2010]. www.joc.com/maritime/k-line-test-ship-scrapping-system. Accessed 18 Apr 2010
Moller-Maersk Group (nd) Our performance. www.maersk.com/Sustainability/Pages/OurPerformance. Accessed 30 Mar 2010
McCarthy L (2012a) Owners' cash claw back blocks green recycling push. Lloyd's List, 8 March 2012, London
McCarthy L (2012b) Calls for China to use shipbuilding facilities for recycling. Lloyd's List, 8 March 2012, London
Messenger B (2013a) Ship recycling levy to undermine competitiveness of EU ports. Marine Insight, 2013. www.marineinsight.com/shipping-news/ship-recycling-levy-to-undermine-com petitiveness-of-EU-ports/18.3.2013.. Accessed 27 Mar 2013
Messenger B (2013b) EU ship recycling levy proposals. Waste Management World, 7 March 2013
Milieu Ltd. & COWI (2009) Study in relationship to options for new initiatives regarding dismantling of ships. Note on the ship dismantling fund. Pros and cons of the three options. Report for the European Commission
Mugdal S, Benito P, Kong MA, Dias D, Cerreño AM (2010) The feasibility of a list of 'green and safe' dismantling facilities and a list of ships likely to go for dismantling. Bio Intelligence Service for the European Commission (DG Env) Final 4 January 2010
Profundo (2013) Financial mechanisms to ensure responsible ship recycling. A research paper prepared for the NGO Shipbreaking Platform, 22 January 2013, Amsterdam
Potočnik J (2011) A more efficient clean up of the ship recycling industry. Speech given by the European Commissioner for Environment at the opening of the photo exhibition 'Broken' organised by the NGO Shipbreaking Platform, 8 February 2011, Brussels
Safety4Sea (2011) A new venture in green scrapping. www.safety4sea.com/pages/4898/3/-a-new-venture-in-green-scrapping. Accessed 31 Mar 2012
Shipping World and Shipbuilder (2012).Ship scrapping to keep Chinese yards in business. March 2012
The Maritime Executive (2011) Siba Ships and Saarland back US$300m green recycling initiative. www.maritime-executive.com/article/siba-ships-and-saarland-back-$330m-green-recycling-ini tiative. Accessed 1 Dec 2011
Wilhelmsen (nd) Green ship recycling. www.wilhelmsen.com/services/maritime/companies/barbership/Pages/greenshiprecycling. Accessed 7 July 2010

Chapter 8
Conclusions

This chapter brings together the various aspects of liability for the safe handling and disposal of hazardous materials encountered during shipbreaking activities, materials whose properties have made severe cumulative and adverse impacts upon those involved in the industry, the surrounding populations and the environment. Following a review of these aspects, it concludes with an examination of the possible way ahead for the international shipbreaking industry in the light of the new *Hong Kong Convention*, with a number of considerations as to how the adverse situation of the industry may, in either the short or medium term, be ameliorated.

To begin with however, it is worth reiterating briefly the personal definitions of liability that have been employed as the thread throughout the study, since legal instruments themselves are usually based upon definitions contained therein. As beneficial recipients of the results of trading, shipowners have both a moral and a legal liability for their end-of-life ships, legal liability usually emanating from a **moral** liability and representing measures designed to diminish or remove the mischiefs caused. The definitions of owners in the newly adopted (IMO) and proposed (EU) legislation now both pointedly include short-term owners (usually cash buyers). Owners' liabilities lie in responsibility for the potential and actual harm emanating from the hazardous materials built into the structures of their vessels and from operational wastes and cargos residues accumulated during the ships' operations and present when the ships finally arrive at the breakers. **Legal** liability begins with design and construction specifications for the ships and ends in removing or minimising the hazards as the ships are dispatched for scrap. Obviously where ships are purchased from other owners, the initial specification liability cannot be transferred, but all other responsibilities transfer with ownership. Some evasion of these liabilities at disposal is common and at times almost inevitable given, on the one hand, the large cash inflows that sale of end-of-life ships represents to their (operating) owners and, on the other hand, the difficulties of finding suitable sites that can accommodate the larger vessels whilst observing the legal requirements of disposal. Avoidance of liability is often achieved through resale or reflagging prior to scrapping and in this owners may be facilitated by the mechanics of anonymity offered by numerous flag states.

M. Galley, *Shipbreaking: Hazards and Liabilities*, DOI 10.1007/978-3-319-04699-0_8,

Similarly, operators of the shipbreaking sites have both a moral and legal duty, firstly to their workers to protect them from the hazards of the relevant materials (as well as an extended responsibility to provide safe working conditions), whilst on a wider sphere protecting all who live in proximity to the sites and to the surrounding ecology. Duty of care legislation also applies on an international, transboundary basis.

The **practical** liabilities recognise the difficulties of stripping if not all then at least the bulk of hazardous materials from a ship whilst leaving it in a seaworthy condition and able to progress under its own power, difficulties that the breakers have often been willing to accommodate since their regard for the consequences of handling such substances may not, in many instances, have been high. Once taking title to a vessel, shipbreakers then proceed with demolition with haste, since the costs of credit and of full preparatory decontamination are high. The proposed EU Regulation does require that owners of EU-flagged tankers sent for demolition ensure at least that cargo tanks and pump rooms are in a condition that is ready for certification as safe-for-entry and safe-for-hot-work.

Ships and shipping operate at a global level that places practical liabilities in the face of other responsibilities if shipbreaking is to continue—as it must. Some 700 ships or more are delivered to the breakers each year and to facilitate delivery the ships need to retain an inherent seaworthy integrity that a complete decontamination prior to departure would deny.

8.1 The Role and Provisions of International Law

Shipbreaking is a heavy, dirty, dangerous and polluting industry and its relationship to any relevant international laws has usually been one of avoidance. Nevertheless, shipbreaking is an industry that appears to be intrinsically sustainable, offering a range of benefits to the host states, yet largely carried out in a manner that appears to be distinctly unsustainable. It is a migrant industry that has followed a line of avoidance to the growth of human and environmental health and safety legislation around the world and, although it may be found operating at some level in most coastal states, it is currently concentrated in China, Turkey, Pakistan, India and Bangladesh, of which India and Bangladesh currently account for most of the large ocean-going ships demolished.

Whilst shipping has grown into a truly global industry in which the design, construction and operation of ships is closely regulated by a large and growing number of international legal instruments, the same cannot be said for the shipbreaking industry *per se*, which historically has been allowed to continue in a somewhat *laissez-faire* style by various host states. Legal instruments directed specifically at shipbreaking have been slim, contentious and easily circumnavigated; the relevance of the *Basel Convention* has been strenuously challenged by most of the shipping and the shipbreaking industries alike. Whilst a number of disputes have arisen over the export/import of specific ships for demolition, disputes between states have not been taken to any formal dispute resolution mechanism; disputes between organizations (NGOs) and states on the other hand have, in a (small) number of instances, resulted in judgements from the courts of the states involved in favour of the NGOs. In little over a decade, however, shipbreaking has been dragged out of the darkness of relative obscurity into the glare of an international treaty, due to the efforts of the IMO and the EU, spurred on at all occasions by the efforts of the campaigning NGOs, led initially by the media-smart Greenpeace and subsequently the NGO Shipbreaking Platform. By these approaches, the question of moral liability raised by the NGOs has been transmogrified into a matter of the legal liabilities involved in shipbreaking, whilst the adoption of legal instruments aimed directly at end-of-life ships recognises also some of the practical liabilities that such an industry generates.

The NGOs' campaign, although meeting with resistance from both the shipping and the shipbreaking industries from the outset, advanced through three parallel streams. On the one hand, it provided a catalyst for changes to international law and judgements from national courts together with changes legislation, especially in Bangladesh, hitherto regarded as one of the worse performing states with regard to shipbreaking. At the same time, campaigning NGOs have also been successful in persuading a number of shipping companies to develop a system whereby ships are scrapped under more controlled and independent supervision—prompted, no doubt, by considerations of corporate social responsibility/public image. From this in turn have arisen new methods of ship disposal, which can lie alongside and be complementary to the new Convention or even evolve in their own right, should the Convention have an extended incubation period before coming fully into force. These emerging practices represent the recognition of a positive opportunity to improve the state of shipbreaking, (as well as commercial opportunities) not hitherto engendered under a basic command and control regime. At the same time. The organization kept up a constant stream of information and campaigning directed at all parties, whether within the industry or without.

Developments in international control have also been the preoccupation of the EU, which is less bound by the consultative approach that the IMO, with its 167 Members, has to adopt, and the EU does not attempt to hide the fact that it seeks to impose conditions or standards that go beyond those required by the Convention. The proposed *SRR* is intended as a stopgap and as an incentive, designed both to promote the use of approved (initially by the EU) shipbreaking sites and as an encouragement to both Member and non-Member States to ratify the *HKC* as soon as possible. In addition, the Regulation includes for the first time provisions which specify quantifiable penalties for Parties failing to observe these requirements through the hitherto standard practice of sale and reflagging of ships immediately prior to their disposal.

8.2 Effectiveness of Legal Instruments

The absence, or at least the absence of enforcement, of labour and environmental laws has proved to be a significant factor in the high turnover of ships that the main breaking states have been able to achieve, and where until recently, regard for environmental protection has proved to be of a low (or even no) priority in some of these states. Prior to 2009, and central to the debate surrounding international regulation of the industry, focus was laid upon the provisions of the *Basel Convention* on transboundary movements of hazardous materials. This Convention was held by some to be highly relevant to the trade in end-of-life ships, a relevance strongly refuted by most of the shipping and shipbreaking industries, and one that led to a distinct polarization of views concerning the industry. One drawback to both the *Basel Convention*, and subsequently to the *WSR* based upon the provisions of *Basel*, has been the ease by which their provisions have been circumvented by sale and reflagging just prior to disposal to the breakers, together with a distinct lack of sites able to accommodate the largest vessels within the acceptable venues. Another drawback has been the lack of responsibilities defined for all below state level, yet the absence of a specific inclusion of end-of-life ships in *Basel* does not necessarily equate to its exclusion or non-applicability. Shipping is quite distinct from other land-based, waste generating industries in that it is somewhat more difficult—and at times even impossible—to define the point at which a ship might become waste, since ships can move freely on the high seas and between different jurisdictions. The decision to dispose may be made at any point and can effectively negate the concept of an exporting and importing state, hence efforts by some states to thwart the requirements of *Basel* with such mobile waste have been easily achieved.

For much of the time that the problems of today's shipbreaking industry have been in the public arena, the arguments as to the relevance of *Basel* have been central; it was only in 2011 that the unenforceable nature of the Convention in its current format was acknowledged by both the EU and by (most of) the Parties to the Basel Convention, since which point the arguments have been able to move forward to the competence of the new *HKC*. Other provisions, which might more easily be defended as being related to shipbreaking, have been limited to a trio of international guidelines, whose observance has been voluntary and, as a consequence, largely ignored.

Internationally, legal provisions directed at shipbreaking have been demonstrated to be ineffective in both the extent to which they have been observed by the breakers and enforced by the regulators of various states. This situation has especially been the case in the sub-continent, where courts, ministries and (local) state authorities appear at times have shown scant regard for each other or for consistency in observing rulings from apex courts. It is concluded therefore that legal regulation in many places has been characterised by a weakness of will to enforce in the face of national interests. States which have in the more recent past shown a stronger governmental control in enforcing health and safety and environmental controls—namely Turkey and particularly China with its strong central government control—have resulted in higher costs for the breakers and hence lower prices for end-of-life ships paid to owners. Nevertheless, these centres have not only survived but are increasingly being regarded as setting a standard for the sub-continent, whilst demonstrating at the same time that some shipowners are willing to internalise costs of environmental protection as a part of their ship disposal programmes.

8.3 Registration, Anonymity and Liability

The decision to examine the role of shipping registers, and in particular certain open registers, arose from the propensity of many owners to choose flags for the final voyages of ships that were much less frequently employed during trading. On change of ownership, a vessel often undergoes a change of identity—ships which are finally sold out of service are often renamed by new owners who change the name of a ship and its registration to an open register that offers short term, low cost registration, as well as low level regulatory enforcement, for the final voyage to the breakers. Again, this is not an illegal practice, but when a vessel's identity is changed—and sometimes more than once—at this stage in a vessel's life, even whilst the ship is at sea and en route to the breakers, it is difficult to consider that this is for reasons of operational economies, but for reasons of obfuscation.

An examination of the provisions offered by certain open registers indicates that as well as the factors referred to above, their attraction for many owners is also the extensive financial measures that they offer to promote the anonymity of owners, the websites of some registers offering equal prominence to the provision of financial services as to maritime services. Such facilities may be very successful in hiding both the identity of the ship and the identity of the owner, together with any inherent liability for what may often be sub-standard ships operated by one-ship companies. Whilst many ships arrive at the breakers under new names, some also arrive with both name and IMO number painted out in an attempt to conceal identity. The fact that many ships arriving at the breakers under flags that they would not carry during their operational lives is an indication of the attraction of these particular registries.

One aspect that will be worthy of observation once the *HKC* comes into force is the impact that the Convention is likely to have on the fortunes of certain of these open registers, a number of whom actually refer to themselves as 'flags of convenience' and operate via third parties with little or no national connection to the state in question, merely taking on the operation of the register as a purely business (*i.e.* cash-generating) venture. The *HKC* is based upon an extensive system of inspections, approvals, permits and notifications that may represent a challenge to certain maritime administrations where they actually perform few maritime functions beyond basic registry. The scope of registries to which shipowners have recourse in the disposal of their vessels may thereby become diminished. Having said that, one should not presume that all the small, open registries are of dubious intent; we are presented with the dichotomy that although the flag of St. Kitts Nevis flies at the stern of numerous vessels *en route* to the breakers and is included in the Paris MoU (2013) black list report for 2012, this state has to date been one of the five states that to date has signed and expressed open support for the new Convention. Nevertheless, a renewed focus upon liabilities as represented by the new Convention may produce a shift in the employment of certain flags at disposal time for owners wishing to observe—and show observance of—the new legislation.

8.4 The Case Studies

The case studies selected relate to specific ships that aroused distinct controversies at the time of their attempted disposal, prompting a number of judgements from various national courts, themselves the result of persistent lobbying by the campaigning NGOs—at times over a period of years for some individual ships. In three of the five cases where judgments were made, the legal reasoning was based upon the relevance of *Basel*, European waste law and ECJ case law upon which it was based. The Indian case law relating to the *Blue Lady*, however, appears in retrospect to be one of justification for scrapping, given the selective disregard that appears to have been paid to much available evidence and questionable findings of physical inspections. The case study of the *Riky* in India also demonstrated the manner in which the confusion and controversy surrounding the relevance of *Basel* has at times enabled national sovereignty to produce a distinctly subjective interpretation. The cases of the *Riky*, *Kaptain Boris*, *Chill* and *Platinum II* demonstrate the inability of the various national environmental organizations to affect the ultimate fates of these vessels once they have left the various jurisdictions.

Although these court judgements represent only a tiny proportion of instances of ship movements being successfully challenged, they did nevertheless keep alive the debate on the relevance of *Basel*. At the same time, the majority of ships continued to arrive at the breakers unencumbered. The Stop Notice issued to the *Margaret Hill* was also ultimately unsuccessful in preventing the ship sailing for demolition, yet this potentially seminal event sent distinct ripples through the shipping industry as the possible precursor of future arrests.

Whilst it is not possible to consider revisions to the *HKC* until the Convention at least comes into force, the question of penalties and/or bonds is worthy of consideration in cases where stated fate and destinations of ships seeking departure permission for 'repairs' appear dubious and in the light of the above cases of the *Platinum II*, the *Chill* and the *Kaptain Boris*. Whilst the *Platinum II* was very much a question for the American Protection Agency, and therefore out of European jurisdiction, it should not be beyond the means of the EU to enact appropriate measures to counter cases of misrepresentation.

8.5 The Hong Kong Convention

Pressured to some extent by the activities of the NGOs directly and upon its Members, the IMO developed and adopted in 2009 the *HKC*, based upon a suite of accompanying voluntary guidelines, which were finalised at the end of 2012. This is the latest of a large number of treaties, guidelines and protocols facilitated by the IMO, and one that was concluded in record time. In common with all the other legal instruments facilitated by the IMO, the *HKC* is not, unfortunately, subject to any enforcement provisions by the Organisation, but is intended to rely largely on the legal standards of operation, inspection and enforcement designed and imposed by the various shipbreaking states over the industry that they wish to devise and subject to whatever national interpretation they deem suitable.

Whilst some parties regard the adoption of the *HKC* as a major advancement in shipbreaking, its format or lack of content on certain issues is considered by others to be a betrayal of the opportunity to impose conditions whereby both the moral and legal obligations of shipowners and shipbreakers with regard to liability could be reinforced. Although the Convention contains requirements relating to ship surveys and the documentation of hazardous materials on board, the focus must remain with the provisions, procedures and performance of the breaking yards themselves since the basic reason behind the movement to improve conditions in many of the sites was the principal driver behind the new Convention.

Perhaps the major criticism currently laid against the new Convention is the challenge that beaching is not to be banned. The fact that beaching can be a hugely polluting activity seems to be beyond doubt and the demands from NGOs that it be stopped immediately and its prohibition incorporated into the *HKC* is understandable, but perhaps not totally practical, at least in the short term, given the realities of the shipping industry and the nature of ships themselves. The inclusion of a proportion of the shipbreaking capacity as one of the conditions for coming into force of the Convention effectively secures the continuation of beaching, at least in the short term, since to outlaw this practice would instantly remove a major share of the world's shipbreaking capacity or effectively prohibit the shipbreaking states of the sub-continent from ratifying the Convention. With regard to the treaty, beaching is an activity that may gradually be discouraged, perhaps as a precondition for new breaking enterprises, given that the level of activity and the number of breakers in action at any one time can wax and wane extensively, dependent upon a range of economic factors in both the commercial and fiscal areas. Once in operation, the scope for progressive enhancements to the Convention is open to the international community, as per the *MARPOL* Convention.

Alongside the dangers from hazardous materials are the risks arising from the actual mechanics of ship demolition and the hazardous conditions under which the workers have traditionally operated—falling from heights, being struck by falling materials and structures, lack of PPE etc. What is needed is a full understanding of the dangers, *plus* a working knowledge of how to remove and dispose of them safely, *plus* the provisions *and* the will to act in an appropriate manner. A formal training of both the workers and the yard owners would seem to be the appropriate pre-requirement, thereafter backed by the incentives to the yard owners of ensuing commercial gain for an approved facility, *and/or* industry/government/public pressure to act accordingly.

A lack of specific enforcement and dispute resolution measures is common with IMO treaties and it may be understandably assumed that interpretation of the requirements of *HKC*'s guidelines is undertaken in a distinctly nationally advantageous manner, yet the Convention is open to a second line of regulation that is somewhat different to its predecessors. The choice by shipowners of shipbreaking sites approved and certificated by national state authorities still has to obtain the approval of the relevant flag states, as is the approval of the recycling plans produced for their ships, whilst shipowners also will need to ensure that their choices are to pass muster in the public eye. In short, operators of shipbreaking sites will be subject to commercial as well as legal pressures, effectively putting their operations under review by both home and flag states jointly. In such circumstances, disputes involving the selection of breakers could be resolved by a simple change of selection.

Whether the perceived lacunae are ultimately addressed or not, the significance of the Convention to date is that it now gives formal international recognition to the problems associated with today's shipbreaking industry, which requires a genuine international effort to remove them. Recognition of equivalence to *Basel* by the EU and by the majority of *Basel* states should provide a major fillip to the new legislation. It also offers an impetus to those new operations and practices that are currently developing in ship disposal in support or in parallel to the Convention.

8.6 Quo Vadis?

So will the shipbreaking industry undergo a massive revision in its operations and practices, with shipowners accepting and acting upon a moral liability for the hazardous materials contained within their end-of-life ships, from which they have received the rewards of beneficial owners, and to the extent that the campaigning NGOs would wish? The answer is most probably to be 'no'—at least in the short run—and although there are a growing number of owners who do recognise such liabilities, the rate of change starts from a slow beginning. For perhaps the first time, both the new *HKC* and the *SRR* place liabilities squarely with the owners of ships, although the extent of the liabilities is not generally detailed. The final and following section therefore considers a number of ways in which the industry and its new Convention might be directed to achieve the results anticipated.

8.6.1 Entry into Force

As a very basic recommendation, the formal coming in to force of the *HKC* should be pursued as a matter of priority, by the application of whatever political pressure the IMO and the EU acting in concert are able to impose, given that the underlying sets of guidelines have been approved since the end of 2012. Particular emphasis should be placed on ratification by the major shipbreaking states, since these are central to the Convention's coming into force. Similarly, the European *SRR* should be implemented, not only as an instrument in its own right, but also as a measure designed to promote action on the Convention. In addition, adoption of the interim recommendations contained in *Selling Ships for Recycling*[1] should be strongly promoted by its supporting organisations such as BIMCO and the ICS, as an aid to accepting, and preparing for, the Convention coming into force and as part of a unified effort of all parties concerned.

8.6.2 Addressing the Question of Beaching

Once the Convention actually enters into force, attention should be given to its perceived lacunae, especially the practice of beaching, which the conditions underlying adoption of the Convention effectively preserve in the short run. Obviously a full or even limited early elimination of beaching is preferable, but establishing such a regime as an *a priori* requirement is most unlikely to gain the support for the Convention by states whose practices are based principally upon this process and whose support is vital. Whether the elimination of beaching later becomes a specific component of the Convention or whether inducements are compiled as per provisions to promote compliance (as appear in other treaties) is a matter of judgement. A direct requirement for at least hard standings might be considered a precondition for new breaking enterprises, given that the number of yards actually in operation changes frequently; already cracks are beginning to develop in the strong defence of the beaching proponents as moves towards some of the provisions and practices of Chinese and Turkish breakers in terms of (partial) hard standings for some of the secondary breaking activities are starting to emerge in the sub-continent and encouragement to promote this move would be a positive measure. Since the provision of such a measure would represent a somewhat significant shift from today's largely low-capitalised industry, some external funding might be sought to assist their construction whilst simultaneously discouraging the current frequency with which operators enter and leave the industry. The call of the Environment Committee of the European Parliament on 26 March 2013 to ban beaching may be regarded as a first step and a fillip to the eventual cessation of beaching.

[1] Maritime International Secretariat (2009).

8.6.3 The Emergence of New Centres

Given the premise that there are likely to be certain flags that are not a Party to the Convention, plus the cost of developing SRFs to the standards defined in the guidelines, there is the possibility of the emergence of new SRFs developing in states where the industry is not yet operational on any large scale. The emergence of such sites needs international collective monitoring by the IMO, given that they may appear to offer alternate (if illegal) scrapping destinations for owners in Party states (especially during periods of sharp downturn for the shipping industry). Once the questionable relevance of the *Basel Convention* has been replaced by measures specific to shipbreaking, then enforcement has a greater chance, especially if the penalties contained within the *SRR* are maintained and even adopted elsewhere.

8.6.4 Pre-cleaning

The controversy of pre-cleaning has run throughout the debate on shipbreaking, although the actual extent and responsibility for this has hitherto never been wholly defined. It would end an ongoing argument for this activity formally to be defined, and in terms that recognise that a thorough and complete removal of all hazardous materials from a ship is likely to preclude its further sailing—hence pre-cleaning should formally be recognised as an initial part of the demolition process, with responsibility lying with the breakers, whilst all efforts to remove or reduce cargo residues might lie with the owners. Regulation 8.2 of the *HKC* merely requires that cargo residues, remaining fuel oil and wastes on board should be removed prior to the ship entering the SRF—the Convention does not currently allocate the responsibility for this action.

Although there is an obvious advantage in ships arriving for demolition in a state as hazard-free as possible, the prior removal of all hazardous materials is one of the major practical difficulties referred to above. Given the cost and potential hazards of towing dead hulks over long distances, apart from the distinct lack of tugs to perform such tows, there is some definite environmental advantage in a ship making its final voyage under its own power and sometimes carrying cargo on its last trip. Furthermore, dead hulks cannot be easily grounded; the hulk of the *Blue Lady* was finally grounded at some distance from the high water mark. The question of precleaning prior to demolition is perhaps not one so much of exactly where this might be performed, but of the provision of adequate facilities to ensure that it is performed with the minimum of risk.

8.6.5 Waste Disposal

The provision of full cleaning and waste disposal facilities at each and every breaker's sites is unlikely to be a practical option, given the intermittent operational/non-operational nature of some sites, together with the physical limitation of actual space allotment. The provision of common (perhaps third-party) facilities for the removal and safe disposal of wastes offers a more realistic alternative and one that can be operated under a more co-ordinated monitoring regime than would be the case of individual enterprises. Suggestions for such facilities were amongst the responses made to an EU consultation on the 2007 *Green Paper*. It is at such a point that preparing a ship ready for hot-work might also be undertaken. Where disposal provisions prove difficult to establish, or as an interim measure, recourse may even be had to repatriating specific hazardous wastes to the country of export? If the breaker does not have the facilities/approval to remove hazardous materials at his yard, then the *HKC* recognises that they may be removed elsewhere. Such removal facilities, however are not necessarily certificated SRFs under the *HKC* and thus should come under the provisions of the *Basel Convention*.

Both the *HKC* and the European *SRR* exclude the actual activity of recycling the material output of the breaking yards, yet the subsequent poor handling and disposal of hazardous wastes beyond the actual breakers' yards is still a prime generator of pollution. Although Regulation 20 of the Convention addresses the subject of hazardous material management, a monitoring regime along the lines of the previous item (with the reporting of wastes included with the provisions for state reporting of demolitions) is necessary to cover this post-demolition aspect of shipbreaking, although it is acknowledged that the requirement for the reporting of hazardous wastes under *Basel* has never been one that has been rigorously observed.

8.6.6 Monitoring Interpretation

One specific danger of controls based upon voluntary interpretation of guidelines is the subjective manner in which treaties in the past have been operated, which in turn is a function of the exercise of national sovereignty. It is recommended that the basis upon which guidelines are interpreted at a national level be communicated to the IMO in order that a comparison be undertaken to ensure that a level playing field is in operation. The results of major discrepancies should be circulated to all Members and should play a role in flag states' approvals of proposed demolitions.

Similarly, the requirement to define and operate national legislation and monitoring standards required by the Convention should also be compared on a nation-by-nation basis—reporting by states of the legal measures that they have defined is already a requirement of the Convention. The claims made by many breakers as to the operation of their sites to international ISO standards, for example, is not one that engenders confidence.

The above two recommendations may differ in their applicability to other treaties in that, with the *HKC*, the successful future of many breaking sites is as likely to depend not only upon the jurisdiction in which they operate, but also depend upon their acceptability to their customers and the maritime administrations under which they in turn operate. In short, shipbreakers effectively will be answerable to their own states on a legislative basis and to other flag states on a commercially acceptable basis. The latter may actually prove to be the more effective, given a strong response from flag states of vessels destined for scrapping. Whilst this may be a matter of little or no importance to some flag states, its relevance amongst EU Member States is likely to be observed in a more robust manner.

The processes of authorisations and documentation required by the *HKC* is likely to provide certain open registries with difficulty in meeting these requirements, given the lack of facilities that such registries are able to offer, and their performance should be monitored by the IMO for further inabilities to meet the basic concepts of the *HKC*. Further, to counter a major function of such registers as providers of anonymity, the OECD proposal to replace the concept of anonymity with one of confidentiality is worthy of reconsideration.

8.6.7 Scope of the HKC

It is recommended that state-owned or operated ships be brought within the scope of the Convention. Whilst other excluded ships (below 500 GT) may collectively amount to a significant volume of shipping scrapped each year, the number of units so covered would be very large and difficult to control. Warships and the like, on the other hand, are relatively few in number, are more likely to be of a greater age than end-of-life commercial ships and contain a higher level of hazardous materials. The fact that they are under state ownership should render the control of their disposal easier than with ships operating under private ownership. Such ships are usually fully deactivated prior to disposal, which does not have to be undertaken in foreign yards.

8.6.8 Technology Transfer

The transfer of appropriate technology should be actively promoted, such transfers most likely relating to the safe handling and disposal of wastes, especially hazardous wastes. The provisions made for the training and monitoring of the (intended) Indian breakers of the *Clémenceau* represented a missed opportunity to form the nucleus of an improved approach to breaking at Alang. Nevertheless, any resultant improvements in practices also need to be accompanied by improvements in the level of legislative control and monitoring if the net result is to prove positive.

8.6.9 Penalties for Non-compliance

The proposed European *SRR* contains detailed penalties for owners who seek to circumvent defined and approved measures for their end-of-life ships. Such provisions represent the first time that specific penalties have been so defined and their use in other regimes is to be encouraged. Whilst again this is a measure that is subject to national sovereignty, the adoption of such formal measures within the national enabling legislation required by the *HKC* would provide an indicator of the degree to which a state recognises the spirit of control that is intended by the new Convention. Article 1 of the *HKC* calls upon IMO Members to cooperate in the compliance and enforcement of the Convention, without specifying how such compliance and enforcement might be defined or facilitated when the IMO has no authorised powers of enforcement.

8.6.10 Reclassification of Shipbreaking Sites

The response of various shipbreaking states to NGO criticism has been to declare their sites to be restricted areas. A positive step toward demonstrating good faith in the acceptance and observance of the Convention would be to de-restrict such sites and render them open to view. Following on from this, it was the monitoring of pollution levels at certain sites in the sub-continent and their independent analysis that gave the edge of verisimilitude to the surveys that were undertaken by Greenpeace. In a reverse manner, therefore, continued pollution sampling, together with the publication of the results, could act as an indicator of the success—or otherwise—with which new guidelines have been effective (whilst bearing in mind the fact that pollutions already accumulated may take an extensive period to diminish).

8.6.11 Dispute Resolution

Whilst the need for more specific dispute resolution provisions beyond the general provisions in the *HKC* may not arise with any frequency since commercial acceptability may prove to be the deciding determinant for a breaker's or even a state's successful shipbreaking activities, it is nevertheless recommended that formal provisions such as those contained within *UNCLOS* for an International Tribunal for the Law of the Sea be included within the *HKC*, in common with other maritime treaties.

8.6.12 Recovery Optimisation

Reference has been made earlier to the fact that much of the scrap metal obtained from ship demolition is regarded in the same manner, *i.e.* as a uniform material representing the lowest common denominator. It would be a boost to sustainability if materials and components of a high grade could be readily identified and used to optimum effect. The international motor industry has come to adopt standard methods of material identification by means of labelling and whilst it is not envisaged that such a system be employed throughout structures as large as ships, a simple listing or diagrammatic representation of the location of high value resources within a ship might form a most useful adjunct to, say, the IHM.

Finally—perhaps it is more of an observation than a recommendation, but it has become apparent during discussions how members of one organisation within the UN can openly contradict the expressions and opinions of another. This does little to engender confidence in an organisation which claims to represent a united voice.

8.7 Postscript

In carrying out a research project, the researcher must be willing to accept a change in opinion or outlook from that which prevailed at the outset, otherwise the research becomes one of simple justification. As a result of the researches undertaken, the author's original viewpoint of full agreement with the sentiments of the NGOs has been moderated; although recognition of the particularly adverse impacts of the industry on both humans and the environment is still present (and is likely to remain so for some considerable time), the practicalities of dispatching ships for demolition also need recognition. A sudden and complete abandonment of the beaching activities is not one that currently is either economically, technically or, in some circumstances perhaps even socially, realistic. Further, and although such a move might have an obvious (if not immediate) positive effect upon the environment in terms of release of contaminants to all mediums, prolonged storage of a rapidly accumulating fleet of end-of-life ships is not an option—witness the growing environmental threats generated by the static reserve fleets in the USA. Shipping breeds end-of-life ships, which breeds shipbreakers. If nothing else, the *Hong Kong Convention* finally demonstrates an international concern over the impacts that the industry has imposed upon its workers and environment and an intent to rectify at least some of its adverse aspects. If liabilities are not to be fully redressed, at least they may be redefined and re-apportioned.

References

Maritime International Secretariat (2009) Selling ships for recycling. Guidelines on transitional measures for shipowners in preparation for the entry into force of the IMO Hong Kong International Convention for the Safe and Environmentally Sound Recycling of Ships. MIS, London

Paris MoU (2013) 2012 annual report on port state control. Paris MoU, Paris

Appendix: Primary Hazardous Materials and Their Properties

Material	Locations	Hazardous properties	Legislation
Asbestos	Thermal and sound insulation. Used as lagging, especially in engine room. Now largely replaced by rockwool.[1]	Airborne particles may remain in suspension for a long time. They accumulate in lungs causing lung cancer, asbestosis and mesothelioma. Symptoms may not show up until many years after exposure.	USA use banned 1989, *EU Directive 87/217/EEC*.[2] Use in ships banned by *SOLAS*, Chapter II amendment 2000.
Polychlorinated biphenyls (PCB)	In a wide range of materials, including cable insulation, transformers, capacitors, switches, some	Carcogenic, bioaccumulate. Associated with cancer, liver, neurological and immune system	Production halted in USA 1979 and in Europe from 1978. Trade regulated by 1998 *Rotterdam*

(continued)

[1] Asbestos containing materials (ACM)—as distinct from simple asbestos—is any material that contains more than 1 % asbestos. ACMs were widely used in the construction of vessels beginning in the 1870s and continuing through the early 1980s. There are three types of ACMs:

- Thermal system insulation; ACM applied to pipes, fittings, boilers, tanks, ducts, or other HVAC (heating, ventilation and air conditioning) components to prevent heat loss or gain.
- Surfacing material; ACM that is sprayed-on, troweled-on, or otherwise applied to surfaces, such as acoustical plaster and fireproofing materials on structural members, or other materials on surfaces for acoustical, decorative, fireproofing, or other purposes.
- Miscellaneous materials; all other ACM, including but not limited to, acoustical ceiling tile, resilient floor tile and coverings, concrete pies, siding, and roofing materials.

FutureNet Group report on the pre-decommissioning asbestos inspection carried out on the Oceanic at San Francisco January 2008, prior to her disposal.

[2] *Directive on the prevention and reduction of environmental pollution by asbestos 87/217/EEC*.

M. Galley, *Shipbreaking: Hazards and Liabilities*, DOI 10.1007/978-3-319-04699-0,

Material	Locations	Hazardous properties	Legislation
	paints, pumps, cranes etc.	damage. Absorbed through skin, inhaled or digested.[3]	*Convention*[4] and *2001 Stockholm Convention.*[5] Defined as hazardous under *Basel Convention 1989.*[6]
Radioactive materials	Low levels in smoke detectors, emergency signs etc.	Ionizing radiation can cause cancer and damage to genetic materials.	
Heavy metal—lead	Paints, cabling, batteries, motors and generators.	Damage to neurological system, hearing, vision, reproductive system, blood vessels, kidneys and heart. Particularly hazardous to children's physical and neurological development.	
Heavy metal—mercury	Light fittings, luminescent lamps, switches, thermometers.	Toxic, bioaccumulative, affects nervous system. Consumption of contaminated fish is an important exposure route.	
Oil and fuel (including hydraulic and lubricating oils, engine oil and grease)	Bunkers, bilges, fuel systems and engine room generally.	Poisonous through inhalation or consumption of contaminated water or fish. Risk of fire and explosion.[7]	
Tributyl tin (TBT) (and other organatins)	Used in anti-fouling paints since 1970s.	Highly toxic to both human and aquatic environments. Skin, eye and	*Convention on control of harmful anti-fouling systems*

(continued)

[3] The common practice of burning wastes containing PCBs and PVC on the beaches releases highly toxic emissions of dioxins and dibenzofuraner.

[4] *1998 Rotterdam Convention on the prior informed consent procedure for certain hazardous chemicals and pesticides in international trade*. Entered into force 24.2.2004.

[5] *2001 Stockholm Convention on persistent organic pollutants* (*POPs*). Entered into force 17.5.2004.

[6] In concentrations of 50 mg/kg or more.

[7] Failure to gas-free ships prior to hot work has caused frequent fires and explosions. Dangerous especially in Bangladesh, where gas-free certification requirements appear to be lax.

Material	Locations	Hazardous properties	Legislation
		lung protection mandatory in industrialized nations.	*for ships*—total ban on use by 2008.
Bilge water	Accumulated stagnant water in bilges, likely to contain a range of pollutants—oil, inorganic salt, heavy metals etc.		
Ballast water	Fresh or salt water taken on board ballast tanks to adjust ship's trim and stability. Quantity may be high when ship sent empty to breakers.	May contain bacteria and viruses, also persistent invasive organisms and sediments.	Convention for the control and management of ships' ballast water and sediments 2004—not yet in force.[8]
Polyvinyl chloride (PVC)	In many materials, including plastics, cable coatings, floor coverings.	May induce cancers, asthma, impairment to human reproduction systems. Burning may generate carbon monoxide, and highly toxic dioxins and furans etc. Burial can release chemicals to groundwater.	
Chloroflurocarbons (CFCs)	Used in refrigerants, fire extinguishing agents, solvents. Becoming obsolete, but still present on many ships.	Non-toxic but ozone depleting substance.	Use in aerosol sprays banned in USA, Canada and Scandinavia in late 1970s. 1985 *Vienna Convention*[9] reduced use of ozone-depleting substances.
Polycyclic aromatic hydrocarbons (PAHs)	Produced during and after torch cut ting when paint smoulders or waste burned.	Many of the 250 compounds are carcinogenic. Cause malignant tumors of the skin, lungs, stomach, intestines and skin. Inhaled as	

(continued)

[8] Numerous treatment systems are currently undergoing development and approval trials.

[9] *1985 Vienna Convention for the protection of the ozone layer* and the *1987 Montreal Protocol*.

Material	Locations	Hazardous properties	Legislation
		fumes or taken up on contact with skin.	
Batteries	Contain heavy metals and sulphuric acid. Associated with radio applications, intercoms, fire alarms, lifeboats.		

Printed by Printforce, the Netherlands